A Great Designer in

THE ORIGIN AND EVOLUTION OF EARTH AND LIFE?

From an Earth Scientist's Perspective

Andreas H. Vassiliou, Ph.D.

CONTENTS

ACKNOWLEDGMENTS

Thanks to my son-in-law Vinnie who provided valuable aid in preparing the manuscript, And to my daughter Katia for designing the original cover.

Also, I am indebted to my friends Dr. Gregory Mestanas and Mr. Rudy Cestone for their Technical and moral support and the numerous fruitful discussions.

INTRODUCTION

We shall march through geologic time, from the earliest earth or the beginning of life on earth to the present. In this time march, we shall analyze scientific evidence obtained from the study of earth materials, primarily **fossils**, and we shall outline **certain key events** in the history of the earth and the origin and evolution of life on earth, **which, we believe, strongly require the participation of a great Designer or Supreme Creator or God**. This is the God or Creator described in most religions.

We shall enumerate and refer to these key events as creation's **Grand Designs**. The number of grand designs could vary and depends on the perspective of the analytic observer. In this discussion, we shall selectively present **7 Grand Designs** which are hereby considered to be **miracles because they are not ascribable to human power or, in other words, humans cannot fully reproduce them or fully understand them or indeed fully explain their creation in terms of the laws of nature.**

In the course of this analytic discussion, we shall also introduce and discuss several other **life milestones**, especially critical evolutionary changes such as the amphibians evolving from certain fish, and life crises such as mass extinctions. These life milestones are considered to be essential parts of the overall grand design of the Creator because they provide **supportive evidence which leads to the selected 7 Grand Designs**.

The aim in this analytic discussion is to show that the selected 7 Grand Designs and the other life milestones, presented here, do undoubtedly require the participation of a Great Designer in order to be seen perhaps as more believable or doable, especially if perceived in the context of "perceptible interruptions of the laws of nature by a supernatural Creator."

At the same time, it should be noted that we do have an abundance of theories and ideas, introduced by earth and life scientists and others, which do not see the necessity of a Great Designer or a Creator in order to justify the existence of the seven grand designs and supportive life milestones presented in this analytic discussion. We shall refer to some of these other theories or ideas in the appropriate sections.

This discussion, although based on scientific evidence, generally geologic and more specifically paleontologic or biologic, is written for the average layman. Partly for this reason, we will not list all general or specific references in the context of the discussion or provide a bibliography list. Nowadays, the reader may obtain a wealth of information on any topic by simply using a search engine on the internet.

At times we shall need to define or explain certain scientific words or terms or ideas that are essential to our understanding of this analytic discussion. Some key explanations will appear parenthetically in the text in the appropriate pages where the terms or ideas are introduced, but all scientific words or terms are also explained in the **Glossary**. However, topics such as **minerals and rocks, fossils, The Geologic Time Scale, evolution, the earth's interior**, and **Plate Tectonics** require a little more than parenthetic explanations in the text or in the glossary. These topics are, therefore, introduced in the "**Background Information**" section of the book. The introductory topics are essentially presented in an outline form that is easier to digest or understand.

It is highly recommended to read the introductory topics in the Background Information section before beginning the analytic discussion in Chapters 1 - 7. However, the reader may choose not to read or digest the Background Information topics first or before beginning the analytic discussion. This is because the Background Information topics are essentially provided as reference information which is specifically referred to and recommended throughout the analytic discussion. Also, many of the Glossary entries refer to the Background Information topics by page number and are expected to aid considerably in making the analytic discussion easier to understand.

CHAPTER 1

THE ORIGIN OF OUR PLANET EARTH

If we look at a recent diagrammatic representation of our Solar System, we see 8 major planets that orbit around the Sun. The 4 planets closest to the sun or the inner planets (Mercury, Venus, Earth, and Mars) are classified as the terrestrial or "earthlike" planets (small, dense, mostly rocky) and the 4 outer planets (Jupiter, Saturn, Uranus, and Neptune) are classified as Jovian planets (large, low density, mostly gaseous with rocky cores). Pluto, for a long time considered to be the 9th planet, was downgraded to a dwarf planet in 2006. Pluto is a snowball of gas, water and rock.

In addition, our Solar System includes over a hundred moons or satellites surrounding the planets, a tremendous number of asteroids, most of which orbit the Sun in a zone between Mars and Jupiter or, in other words, between the terrestrial and jovian planets. Our Solar System also includes millions of comets and meteorites, and tremendous amounts of interplanetary dust and gases. Our Solar System is a tiny part of our galaxy, the Milky Way, which is estimated to contain over 200 billion stars like our Sun. In turn, our galaxy is a tiny part of the universe which contains billions of other galaxies.

Today astronomers and other scientists believe that the universe originated about 14 billion years ago in what is popularly called the **Big Bang**. In simple terms, the Big Bang theory states that our universe began with the cosmic explosion of an infinitely small and inconceivably dense and hot point in which all matter and energy were compacted, and both time and space were set at zero. Therefore, there is no "before the Big Bang," only what happened after the big explosion.

What is the most fundamental scientific evidence that supports the Big Bang theory? Two phenomena: **One**, the universe is expanding, an expected condition after a major explosion. Astronomers observe that, everywhere in the universe, galaxies are moving away from each other at tremendous speeds. **Two**, the universe is permeated with a pervasive background radiation. Arno Penzias and Robert Wilson won the Nobel Prize primarily because of their 1965 discovery of the background radiation thought to be the fading afterglow of the Big Bang.

Cosmologists (the scientists who study the origin, the evolution, and nature of the universe) and theoretical physicists cannot describe or explain what it was like at **time zero** of the Big Bang because they do not understand the physics of matter and energy under such incredibly extreme conditions.

The what, how, and why of what happened **after** the Big Bang is not any easier to describe or explain – we face a complicated universe around us. Three papers published independently in 1964 described an energy field, present throughout space, which is responsible for the mass and diversity of the particles in the universe. This energy field or energy ocean is now known as the **Higgs field**. One interpretation suggests that without the Higgs field, all elementary particles would have no mass and, therefore, there would be no atoms and no worlds and no life and no us. The theory caused a four-decade search for a telltale particle known as the **Higgs boson** or the **God particle**. The discovery of the God particle was finally realized in 2012 at the Large Hadron Collider at CERN in Switzerland. Two of the scientists who came up with the Higgs field theory in 1964, Peter Higgs and Francois Englert, received the Nobel Prize in Physics in 2013.

Another interesting theory called **supersymmetry** keeps many a physicist on the search for a telltale particle called **gravitino**. The theory proposes that each of the 17 particles already identified has a nearly identical cousin still hidden in the shadows of the universe. These "missing" supersymmetrical particles are the puzzle pieces needed to solve mysteries such as: How does gravity work? What came before the Big Bang? The search goes on.

All of the above represent the quintessential cry for the Great Designer or Creator not only because we do not understand conditions at time zero but also because we know and understand little of what happened a split second or billions of years after time began. We refer to the creation of the universe as **Grand Design #1**.

The **universe** has been expanding for the last 14 billion years. We shall focus on the last 4.6 billion years of this vast expanse, when our **solar system** (its members described in the beginning of this chapter) formed and evolved. Earth scientists look to the origin and formation of the solar system in order to understand the origin and formation of the earth.

Several scientific and non-scientific theories for the origin of the Solar System have been presented during the last 300 years. Today, most scientists believe that the **solar nebula theory** provides the best explanation that best fits the known facts. Basically, it involves the condensation and collapse of a rotating cloud (nebula) of gas and fine dust. Looking at the process stepwise, this interstellar material (in a spiral arm of the Milky Way galaxy), this huge rotating cloud of gas and dust, contracts (due to the force of gravity) and flattens (due to rotation) and forms a disc where the matter begins to drift toward the center (due to the pull of gravity) to form the Sun. Although most of the matter in the original cloud was concentrated in the Sun, a disc of gas and dust (the solar nebula) remained to envelop it. This enveloping disc of gas and dust formed grains that collided and clumped together (due to gravitational attraction) into small chunks called planetesimals which eventually built the planets by multiple collisions and accretion.

During the last two decades, data gathered by spacecraft carrying planetary probes was expected to provide evidence on the origin of our solar system by comparing the data on the nature and evolution of the planets and other bodies making up our solar system. One surprising discovery: No two bodies in the Solar System are the same. This makes our earth unique from this point of view. Also, as we shall see in the next chapter, the Earth is the only likely planet to be able to support life as we know it.

Chapter Highlights and Conclusions:

1.   The Big Bang is the generally accepted theory for the origin of the Universe but scientists still cannot understand conditions at time zero (before the Big Bang). Also, scientists cannot adequately understand or explain what happened a split second before or billions of years after the Big Bang.

2.   Data gathered by spacecraft carrying planetary probes indicate that no two bodies in our Solar System are the same. This makes the earth unique in terms of supporting life as we know it.

3.   The introduction of a Great Designer or God allows us to understand or explain the origin of the Universe and the earth's uniqueness for supporting life as we know it. The understanding comes by declaring these events as perceptible interruptions of the laws of nature by a supernatural Creator. In our analytic discussion, this is what we designate as <u>Grand Design #1</u>.

CHAPTER 2

WHY LIFE ON EARTH ONLY?

Thus far, we found life, as we know it, on earth only. Simple organic compounds or amino acids, presumably the basic requirements for life formation, were reportedly found in a couple of meteorites that fell on earth.

As pointed out in Chapter 1 above, our Solar System is a tiny part of our galaxy, the Milky Way, which is estimated to contain over 200 billion stars like our Sun. If we were to plot, to scale, the area of our Solar System on a map of our galaxy, the Milky Way, we could hardly place it as a visible spot on the map. If we wanted to do the same for our galaxy on a map of the universe, our galaxy would appear as a tiny smear. Do these billions of stars in our galaxy and the trillions in the universe have planets orbiting around them as our own star, the Sun? We expect at least some of them to have planets orbiting around them.

Thus far, after many years of astronomical quest in our galaxy and other areas of the universe, astronomers only found a handful of planets that appear to be probable candidates for life of one kind or another. Over all, the world's astronomers now know of almost 1000 planets outside our solar system, but strongly believe that our galaxy contains billions of earth-like planets and expect to find what they call **Earth 2.0**, the "lost twin" of our earth.

The most recent analysis of data from NASA's Kepler spacecraft (Erik Petigura, Proceedings of the National Academy of Sciences, November 4, 2013) reported that there could be as many as 40 billion habitable earth-size planets in our Milky Way galaxy. Other studies in early 2013 found that 15 to 50 percent of the smaller, numerous Red Dwarfs in our galaxy have Earth-like planets in their habitable zones (not too hot, not too cold). How are these planets discovered in distant constellations of our galaxy? When orbiting planets pass in front of stars they cause dips in the brightness of these stars. The Kepler spacecraft was launched in 2009 to look for these planets by looking for dips in star brightness.

Although these newly discovered planets are referred to as Earth-size or Earth-like, nobody knows what their masses are (rocky like the Earth or balls of ice or gas) or whether life can exist on them. Astronomers did not as yet discover any planets that are exact analogues of the Earth in terms of all three of the following critical characteristics: size, orbit type, and star type.

Some astronomers also look for life on the moons that orbit the Jovian planets listed above. They believe that there is water beneath the thick ice covering these moons, liquid water where life originates. But this does not mean that life does exist on these planets or moons for sure. Can this be the rule of the endless universe that life exists only on a super-tiny and relatively insignificant spot known as the earth? In general, astronomers believe that this cannot be so, and the quest to find life elsewhere continues.

What makes the earth the ideal planet where life can exist? Several factors have been introduced through the years that propose to make the earth unique in terms of supporting life. In reality, the following two characteristics of the earth cover all factors. The first is that the earth has the **right temperature**. This temperature, related to the Earth's distance from the Sun, is such that most of the water on the planet is in the liquid state. This is important because life, on the basis of fossils found, originated in liquid water. It is evident, therefore, that liquid water is essential to life. The second factor is that the earth has the **right size**. If the earth were smaller, its force of gravity (which is directly proportional to its mass) would be smaller or weaker and the atmosphere (which is needed by practically all organisms) would escape into space. If the earth were larger, its force of gravity would be stronger and it would pull the atmosphere too close to the earth's surface thus suffocating life on earth

At this point, since we did not find life to exist in any other part of the vast universe, we may conclude that the earth and life on earth were designed by a supreme designer to be unique. Of course there are many scientists and non-scientists that support the premise that both the earth and life on earth came to be perhaps by accident or by the normal workings of the laws of nature effecting changes through a vast amount of time which is measured in billions of years. Let's analyze this.

Geologists describe the earth as dynamic because it is constantly moving and changing. Geology, the study of the earth as the Greek word implies, is actually the study of the changes the earth goes through, both past and present. On the basis of this, let us look at the two factors that make the earth suitable for life.

First, let us consider the size of the earth. Did its size change since its origin, 4.6 billion years ago? There is no geologic evidence that the original mass of the earth changed to any significant degree since its origin. There is evidence shown by the theory of Plate Tectonics (see Background Information, p. 79-84) that the shape and size and geographic distribution of the continents and oceans have changed but the total mass remained about the same. So, one may say that the original design of the right size was never modified.

What about the factor of the right temperature? Earth scientists think that the earth at first was probably cool and of uniform composition, mostly silicate minerals or compounds of Si (silicon) and O (oxygen) plus other minerals of Fe (iron) and Mg (magnesium), and of uniform density throughout. Later on, this early cool earth was subjected to showers of meteorite impacts for millions of years. The heat from these impacts plus additional heat from radioactive decay (from radioactive isotopes within the earth) and heat from gravitational compression increased the earth's temperature. This higher temperature caused the melting of iron and nickel which, due to their higher density, gravitationally settled towards the earth's interior and formed the inner and outer core of the earth. The other rock materials formed concentric layers of differing composition and density above the metallic core, producing the mantle of denser Fe and Mg silicates above the outer core and the lighter crust as the outermost layer of lighter silicates above the mantle (see Earth's Interior in the Background Information, p. 76-78).

The above events were responsible in creating a differentiated or layered planet. This is most significant because this differentiation led to the creation of the crust and the continents. This differentiation is also believed to have caused the emission of gases from the earth's interior (known as "outgassing" from volcanoes or other fractures in the crust) which led to the formation of the atmosphere and the oceans (water vapor condensing on the surface). The metallic core, especially the liquid outer core, is also believed to have contributed in the creation of the earth's magnetic field which helped retain the gases and form an atmosphere.

The above information shows that the right temperature and consequently the ocean water and the atmosphere, all needed for the origin or sustention of life on earth, came into existence a few million years after the origin of the earth, a relatively short time interval when we consider the vastness of geologic time. Does this change the proposition of a supreme designer? Not at all, because as we shall see throughout this discussion, it appears that many of the miraculous designs relating to the origin and evolution of life were introduced gradually, very much like the harmonious movements of a symphony. The design and building of a city is offered as being analogous. An architect designs and builds a public building or a street or a park, one at a time, as part of a single theme, before he or she creates the completed metropolis.

It is apparent to this observer that a **great designer** put together all necessary requirements, which fit together or complement each other like the spokes of a wheel, to make the earth **unique for life as we know it**. This designed uniqueness is considered **Grand Design #2**. Many people, scientists and non-scientists, think that it would be an incredibly exhilarating event if life is found outside the earth. It is also believed that it would be incredibly exhilarating if life exists only on earth, making our earth unique as noted here.

Chapter Highlights and Conclusions:

1.   **Evidence shows that the planet Earth has the right temperature and the right size, two critical factors that enabled life, as we know it, to originate and be sustained on the planet.**

2.   **Planet Earth appears to be unique in our Solar System and the Universe in terms of having the right characteristics to support life.**

3.   **The Earth's uniqueness for supporting life is attributed to the Great Designer or Creator. This incredible uniqueness, within a universe of apparently limitless size and possibilities, is considered a miracle, and is designated as <u>Grand Design #2</u>.**

CHAPTER 3

HOW LIFE ORIGINATED?

Scientists believe that life originated from nonliving matter, not at once but in several small steps, and refer to this process as **abiogenesis**. To make the distinction between living and nonliving, which is not always clear, biologists use two criteria: **(a)** A living organism must have the capacity for self-regulation or, in other words, use raw materials (food) in order to sustain internal chemical reactions (metabolism), and **(b)** a living organism must have the capacity for self-replication (produce offspring).

The above two criteria relate to the fact that all organisms are cellular (unicellular or multicellular), and the cell includes apparatuses which facilitate chemical reactions and also includes the RNA / DNA molecule or the blueprint for reproduction. In addition, all organisms, from bacteria to complex animals and plants, are made up of the same four elements: **Carbon (C), Hydrogen (H), Oxygen (O), and Nitrogen (N).**

Scientists also believe that the following three items are essential requirements for the origin of life: **One**, a source of the necessary four elements that make-up all organisms. In the Archean Eon (see Geologic Time Scale, p. 71) where life originated on the basis of fossils found, the early atmosphere did contain the necessary elements in the gaseous molecules of CO_2, H_2O, CH_4, and NH_3. **Two**, an energy source necessary to promote chemical reactions between the 4 elements is needed. Two such energy sources were available in the early Archean: Lightning and Ultra Violet (UV) radiation from the Sun. The ozone layer that inhibits UV radiation from reaching the earth's surface was not in place at the time. **Three**, the right environment was needed, in this case an atmosphere that contained little or no free oxygen or molecular oxygen (O_2). Free oxygen would inhibit the formation of simple organic molecules or it would destroy any early organic compounds that would form.

(a) Hypothetical – Based on Experimental Evidence

With the three essential requirements for the origin of life in mind, as outlined above, scientists came up with three stages in life formation. These stages are, of course, hypothetical but they are supported by experimental results.

Stage 1. The formation of monomers (amino acids or simple organic molecules): Elements (atmosphere) + Energy (lightning) = Monomers (amino acids)
In the 1950's, Stanley Miller and his co-workers synthesized amino acids in closed glass vessels containing atmospheric elements that were energized with an electric spark.

Stage 2. The formation of polymers (proteinoids): Monomers + monomers = polymers.

Experimenters showed that dehydration and heating of monomers would cause them to link and form polymers or proteinoids. The latter would then spontaneously aggregate into microspheres and grow and divide in a nonbiologic manner.

Stage 3. The formation of a replication mechanism, RNA/DNA nucleic acids.

Organisms require these nucleic acids for reproduction (RNA by bacteria and DNA by higher organisms). **But these nucleic acids cannot replicate without protein enzymes, yet these enzymes cannot be made without nucleic acids.** Recently researchers discovered that some small RNA molecules can reproduce without enzymes. This suggests that the first replicating system might have been an RNA molecule. But the great mystery still remains: **How RNA and later DNA were naturally synthesized under conditions that existed on early Earth?**

This mystery is resolved very quickly if we introduce a Great Designer. This Designer could synthesize the nucleic acids and the enzymes at the same time. To the writer, this represents a miracle in the origin and evolution of life, a grand design event; however, since we are dealing with hypothetical or experimental stages in life formation in this section, we shall designate the event of the DNA synthesis as **one of the Creator's grand designs (Grand Design #4)** when we introduce the actual discovery on Earth of an organism with the first DNA cell, in the upcoming chapter 4.

There are other ideas or theories concerning the origin of life. We shall take a quick look at two of them.

The first one postulates that perhaps life began at **hydrothermal vent systems**. These "warm water systems" were first seen in 1979 in the eastern Pacific Ocean when scientists descended 2,500 meters. Since then, they were seen in all major oceans near mid-ocean ridges or spreading centers (see Plate Tectonics, p.79). At these centers, cold water seeps through the crust and is heated by the hot rocks at depth before it rises and discharges into the sea water as plums of hot water as hot as 400°C. Most of these plums are black (known as **black smokers**) because of dissolved minerals.

The necessary elements (C, H, N, O) for life are also present in sea water. Heat from the hydrothermal vents is believed to have been the energy that caused the formation of monomers (amino acids) through the combination of the four elements. Amino acids have been detected in some of these hydrothermal vents. Those who endorse this view of the origin of life suggest that the next step after the formation of monomers or amino acids is polymerization which is believed to have taken place on the surface of clay minerals found here. In the third and final step, protocells (RNA/DNA) were deposited on the seafloor. It has been observed by scientists manning the submersible Alvin, that life grows incredibly fast at these vents. Tiny warms and sulfide mineral deposits grew extremely large in an interval of three years – from a few inches to several feet.

The second one, a far-flung possibility for the origin of life, was introduced by Dr. Steven Benner, an organic chemist thought to be one of the world's leading experts on the origin of life. He introduced the idea through a lecture he delivered at a geology conference in Florence in August of 2013. He postulated that the planet Mars might have been a favorable place for the first self-replicating molecules (RNA/DNA) to form and, therefore, for life to start. (It was noted above, that the formation of the self-replicating molecules is the 3rd and final step in the origin of life.)

Dr. Brenner and his associates noted that the Earth's early environment, unlike Mars, lacked the necessary minerals such as borates and molybdates that acted as catalysts for the reaction of the original organic compounds to form RNA/DNA. Because of the latter, life formed on Mars first and then was introduced to the Earth. This introduction of life on earth is postulated to have been accomplished through a giant impact on Mars, causing life-laden rocks to fall on Earth and begin life here.

Section Highlights and Conclusions:

1.   It is generally believed that the essential requirements for life to originate on the surface of the earth are represented by the following 3 items: A source of the necessary 4 elements (C,H,O,N) that make up all organisms, an energy source to promote chemical reactions between the 4 elements, and the right environment that would allow the early organic compounds to form and continue to exist.

2.   It has been determined that the early Earth in the Archean Eon did satisfy the above essential requirements: The Earth's atmosphere did contain the 4 elements, lightning and UV radiation were available as sources of energy, and the early atmosphere did provide the right environment because it was practically free of molecular oxygen (O_2).

3.   Having the above in mind, scientists came up with 3 hypothetical stages in the formation of life as we know it, and have been trying to reproduce life in the laboratory by following these 3 stages. The first stage calls for the production of amino acids or simple organic compounds. This has been done in the laboratory by using closed vessels which contain the required atmospheric elements which are energized with an electric spark. The second stage calls for the formation of polymers (proteinoids) by combining the amino acids or monomers. This stage was also accomplished in the laboratory through the heating and dehydration of the monomers. The third stage calls for the formation of a replication system, namely the nucleic acids of RNA (in bacteria) and DNA (in all other organisms). This stage has not as yet been fully understood or duplicated in the laboratory.

4.   No matter which idea or theory concerning the origin of life we analyze, the end result is that life's replication mechanism (RNA or DNA) remains a mystery in terms of how it was naturally synthesized on early Earth. For instance, RNA/DNA nucleic acids cannot replicate without certain protein enzymes, and these protein enzymes cannot be made without the RNA/DNA nucleic acids. This calls for the Great Designer to come to the rescue. We shall designate this rescue as part of Grand Design #4 which represents the first appearance on earth of the DNA molecule (see discussion in Chapter 4).

(b)   The Archean Eon – The Earliest Fossils Found on Earth

At this point, we begin our march through time. We begin with the earliest period in the earth's history, the Archean Eon whose time span is about 2.1 billion years. Together with the Proterozoic Eon, the two eons make up the Precambrian Era (see the Geologic Time Scale, p. 71). For a long time or until the first half of the 20th century, geologists thought that life did not exist in the Precambrian Era which makes up 87% of geologic time. We now have fossil evidence that life at its simplest, mostly bacteria, did exist in the Precambrian Era.

The early Archean environment was generally inhospitable to any kind of life. It was barren, waterless, was subjected to numerous meteorite impacts and UV radiation (no ozone layer), and it had no atmosphere and no burning (no free oxygen or O_2). About a billion years into the Archean, small continents, known as cratons, developed together with oceans and an early atmosphere, both of the latter the result of outgassing of volcanic gases. Molecular or free oxygen (O_2) started to accumulate in the atmosphere, mostly due to photosynthesis by autotrophic (self-feeding) bacteria, until it became 1% of today's oxygen in the atmosphere, by the end of the Archean.

Energy sources such as lightning and UV radiation acted on chemical elements found mostly in the atmosphere, producing the first living things, unicellular prokaryotic bacteria which lacked a cell nucleus and nucleic acids (RNA) which serve as a blueprint for reproduction. All known Archean fossils represent prokaryotic bacteria.

Geologists have found fossils in Archean rocks in Australia as old as 3.5 billion years. The key word here is "found" because life may have existed earlier and we haven't found it yet. Nevertheless, according to this finding, life on earth appears to have originated over a billion years after the earth came into existence. These earliest fossils are known as **stromatolites**, organic, layered and dome-like structures built by **cyanobacteria** or blue-green algae. The cyanobacteria organisms are classified as single-celled, **anaerobic** (they do not require free oxygen), **autotrophic** (photosynthesizing) prokaryots which include bacteria and cyanobacteria. Of course, we can speculate that the simpler **heterotrophic** or non-photosynthesizing prokaryotes must have existed first, since photosynthesis is a relatively complex metabolic process. As of now, the simpler heterotrophic bacteria were never found in the Archean.

Nevertheless, it is evident that the most significant event in the life history of the Archean was the development of photosynthesis, not only because it allowed the autotrophic cyanobacteria to create its own food but also because the process produced free oxygen as a byproduct. Here we like to refer to this event of the development of photosynthesis, a metabolic process too complex for humans to reproduce, as **Grand Design #3**.

So, it appears here that a master design was in effect and evident throughout the Archean, a design that would eventually produce the more complex or higher life on earth. First, the designer created an environment that did not include free oxygen (O_2) so that simple organic compounds and eventually simple life (bacteria) can develop. Then, the bacteria acquired the self-feeding process of photosynthesis that also produced free oxygen that eventually (in the Proterozoic) allowed the more sophisticated aerobic or oxygen-requiring organisms (the eukaryotes) to come into existence.

Section Highlights and Conclusions:

1.   The earliest known Archean fossils represent single-celled, anaerobic and photosynthesizing prokaryotic bacteria (cyanobacteria). The fossils are known as stromatolites (layered and dome-like structures built by cyanobacteria) and were found in 3.5 billion year old Archean rocks in Australia.

2.   In the Archean, the Great Designer first created an environment that did not include free oxygen (O_2) so that simple prokaryotic life (bacteria) could develop and eventually acquire the self-feeding process of photosynthesis, a metabolic process too complex for humans to reproduce. This chapter in the history of life is hereby designated as <u>Grand Design #3</u>.

CHAPTER 4

PROTEROZOIC – THE FIRST EUKARYOTIC (DNA) CELLS

A quick look at our Geologic Time Scale (see Background Information, p. 71) would show us that the Archean Eon lasted for a little over 2 billion years and the Proterozoic Eon for very close to 2 billion years. By the end of the Archean, the free oxygen (O_2) content in the atmosphere reached 1% of today's levels primarily due to the photosynthesizing anaerobic prokaryotes (cyanobacteria). By the end of the Proterozoic, the free oxygen (O_2) content in the atmosphere reached 10% of today's levels because, during this eon, the photosynthesizing aerobic eukaryotes (Archaea) were introduced. One of the main reasons for the faster build-up of O_2 was the fact that the eukaryotes possessed a cell nucleus and reproduced sexually, thus producing more offspring and, therefore, more oxygen

The generally accepted first discovery of a unicellular aerobic eukaryote is the 1.4 billion year old fossils found in the Beck Spring Dolomite of Southern California. Another find, in the Negaunee Iron Formation of Michigan, suggests that the oldest discovered fossil of a eukaryote (a single-celled bacterium, probably some kind of algae) is as old as 2.1 billion years. These eukaryotic cells of the Proterozoic were much larger than the prokaryotic cells of the Archean; they had a nucleus containing the genetic material DNA or the blueprint for reproduction, and, in general, reproduced sexually.

With the increase of free oxygen in the atmosphere, primarily due to the unicellular aerobic eukaryotes discussed above, the **multicellular** aerobic eukaryotes came into existence. These multicelled organisms are not only composed of many cells (often millions) but also have specialized cells (organs) that perform specific functions such as respiration (lungs). The oldest known such multicellular eukaryotes came from 1.2 billion year old rocks in Canada. Some slightly younger similar fossils (around 1 billion years old) were found in India and China. These multicellular eukaryotes reproduced sexually and are members of the kingdom of **Plants**.

The Ediacaran Faunas of Australia represent a unique assemblage of multicelled organisms, the earliest members of the kingdom of **Animals** found as fossils. These were soft-bodied animals preserved as molds and casts in sandstone. These first animals existed during the very end of the Proterozoic, between 670 and 570 million years ago. Fossils of these first animals were recently found in most of the other continents. All of these fossils seem to represent soft-bodied animals but some chitin or calcareous skeletal elements appear to have been present, suggesting the beginning of the process that eventually produced skeletonized animals.

This first appearance of the eukaryotic cell (with its DNA nucleus) marks one of the most important events in the history of life because all forms of life (except bacteria) are eukaryotic. This event of the introduction of the eukaryotic cell on Earth is actually **everybody's birthday!** It is here designated as **Grand Design #4**. As noted earlier in Chapter 3(a), the **synthesis** of the RNA/DNA molecules, discussed in connection with the experimental and hypothetical third stage in life formation, is essentially part of this designation of Grand Design #4 because it points out to the fact that humans cannot come even close to synthesizing these DNA molecules.

The fossil record does not reveal how a eukaryotic cell emerged or developed from an earlier prokaryotic cell. How does science, then, explain the appearance or the evolution of a eukaryotic cell from a prokaryotic cell? A popular theory suggests a kind of symbiotic relationship among prokaryotes is responsible. In other words, several prokaryotes living together, for a long period of time, would eventually become eukaryotes. Of course, mainly because of the long time requirement, scientists cannot reproduce these symbiotic results in the laboratory. Similarly, fossils do not reveal how multicelled organisms originated or developed from the earlier unicelled organisms. Here again, scientists cannot reproduce the process but provide only clues, based on studies of present-day organisms, as to how the transition from unicellular to multicellular materialized.

Chapter Highlights and Conclusions:

1.   The first appearance of the eukaryotic cell with a DNA nucleus, about 1.4 billion years ago, represents the most important event in the history of life because all forms of life (except bacteria) are eukaryotic.

2.   Scientists and non-scientists recognize the vast importance of the appearance or development of the unicellular eukaryotes and eventually the multicellular eukaryotes. Hypotheses abound as to how these eukaryotic cells developed, but nobody can actually reproduce the process or processes that developed these cells. Here, again, these events or developments become more acceptable and more understandable if we assign their creation to a Great Designer. From this perspective, the first appearance on Earth of the eukaryotic unicellular and multicellular cell is designated as <u>Grand Design #4.</u>

CHAPTER 5

THE PALEOZOIC ERA

The Paleozoic (ancient life) is 325 million years (see Geologic Time Scale, p. 71) of interconnected geologic and biologic events. Some of the key highlights of the tremendous biologic change that took place during this era include the appearance of skeletonized animals (invertebrates with exoskeleton or shell), the appearance of vertebrates (fish) which eventually evolved into amphibians which then evolved into reptiles, the first true land animals. Fossils show that plants invaded the land and became the first true land dwellers before animals (the amphibians and the reptiles) did. Major extinctions effecting all Paleozoic animals (both sea and land dwellers) took place at the end of the era. These extinctions represent **the worst life crisis on earth**.

The geologic events that had a profound effect on life development and evolution or diversification relate to **plate tectonics** (see Background Information, p. 79), the generally accepted geologic theory that explains continental drift and sea-floor spreading and associate processes such as volcanism and earthquakes. Examples of plate tectonic geologic events that effected life and evolution include the opening and closing of oceans, the transgressions and regressions of seas over land, and the changing positions of the continents, including the formation of the continuous landmass **Pangea** (see Background Information, p. 81) at the end of the Paleozoic.

For convenience and proper emphasis, the Paleozoic is subdivided into three sections: (a) the Early Paleozoic; (b) the Middle Paleozoic; and (c) the Late Paleozoic.

(a) The Early Paleozoic Era

The Early Paleozoic in this discussion represents the Cambrian and Ordovician Periods, with time duration of 132 million years.

Animals with exoskeletons or shells (invertebrates) appeared rather abruptly at the beginning of the Cambrian. This sudden appearance of multicelled organisms with hard parts is often referred to as the "**Cambrian explosion.**" Because we did not find these organisms as fossils before the Cambrian, we may assume that they had a long history during which they lacked hard parts. On the other hand, perhaps these organisms did exist in the Precambrian but we never found their fossils. The Middle Cambrian Burgess Shale of British Columbia provided one of the best well-preserved group of Cambrian fossils in the world (discovered by Walcott in 1909). Of the group's 107 fossils, 64 represent well-preserved soft-bodied organisms.

The exoskeletons or hard parts provided a lot of advantages such as support for the muscles of these organisms, allowing them to grow larger and have enhanced locomotion. Perhaps more importantly, these hard parts provided much needed protection from predators, UV radiation from the sun, and protection from desiccation caused mainly by regressions of seas. Some scientists believe that these hard parts probably evolved due to various factors, both biologic and geologic, instead of one single factor or cause. In any case, this evolutionary explosion remains a kind of miracle that can be easily explained or understood as the work of a supreme Designer or Creator. Here it is considered to be a **life milestone supportive of the selected seven grand designs**.

The Upper Cambrian also provided fossils of the earliest known vertebrates, armored and jawless marine fish known as Ostracoderms (bony skin). These early vertebrates are classified as a sub-phylum of the chordates whose earliest members were soft-bodied. This suggests that soft-bodied chordates evolved into the hard- spine vertebrates. As usual, the intermediary change is not well documented and the overall mechanism of evolutionary change represents conjecture unless we introduce a great designer or refer to this change as a **life milestone supportive of the selected seven grand designs**.

The Ordovician Period registers great diversification of the members of both the invertebrate and the vertebrate communities. The shelly invertebrates dominated but we also see large-scale reef building by corals. The end of the Ordovician brought major extinctions of primarily shelly invertebrates.

Looking at an area represented by North America and the Atlantic ocean, these extinctions may be attributed, to a large extent, to the geologic event of the Taconic orogeny, a mountain building event which took place towards the end of the Ordovician. As part of the period's plate tectonics activity, it represents the near closing of the Iapetus Ocean (the name given for the earlier ocean in the area of the present Atlantic Ocean) or, in other words, the subduction of the Iapetus Ocean under Laurentia (the name given for the earlier North America). As an example of the interconnected geologic and biologic events, the gradual closing of the Iapetus Ocean would affect marine organisms in terms of limiting their living space.

Is this extinction event part of the great design? It appears to be so because it provided an opportunity for the reef-building corals to dominate in the next period and, as a result, also allowed the eventual great diversification and dominance of the vertebrate fish from which subsequently evolved the amphibians and then the reptiles. We refer to this extinction event as a **life milestone supportive of the selected seven grand designs.**

(b) The Middle Paleozoic Era

The Middle Paleozoic in this discussion represents the Silurian and Devonian Periods, with time duration of 78 million years.

In terms of invertebrates, the Silurian saw the dominance of corals which produced major reef-building in the oceans. The vertebrates, the armored jawless fish (the Ostracoderms), became quite common but we also see the introduction of jawed fish, the non-marine Acanthodians and the marine Placoderms.

Based on comparative anatomy, scientists refer to the appearance of jawed fish as **"evolutionary opportunism."** They believe that the jaw evolved from the modification of anterior gill arches that were part of the jawless fish. This may explain the appearance of the marine Placoderms evolving from the marine Ostracoderms, but the idea does not fully explain the essentially simultaneous appearance of the non-marine Acanthodians. This is another **life milestone supportive of the selected seven grand designs.**

The earliest non-disputed land plants were found in the Silurian. These were the **seedless vascular** plants that lived in a low, wet marshy, fresh water environment. The earliest of these vascular plants were small, leafless, rootless, and, of course, seedless. Slowly they diversified and developed roots and leaves. Their reproductive cycle (in a marshy environment) involved the production of spores which germinated and grew into small plants (gametophyte) that produced sperm and eggs. In turn, the fertilized eggs grew into the spore-producing mature plants (sporophyte).

Terrestrial vascular plants are believed to have evolved from some type of (marine) green algae with the same type of reproductive cycle. The evolution of vascular tissue provided the plant with additional support and allowed food and water to be transported throughout the plant. In making the transition from water to land, the early plants, and later the early land animals, had to overcome problems such as gravity, atmosphere, desiccation, and reproduction methods. Nevertheless, plants slowly developed some of the major structural features which characterize today's plants.

Can we see the hand of a great designer in the introduction of land plants? Yes, without any doubt, because it is difficult to understand or explain the transition from life under water to life under the atmosphere. We may call it evolution from marine green algae to land plants but we still cannot fully explain it, we only label it. We designate this event as another **life milestone supportive of the selected seven grand designs**.

In the Devonian, the **seedless vascular** plants diversified to a large extent. Plants, now with roots and leaves and up to 40 feet tall, produced extensive forests. Other developments in this period included the first land insects (flightless) and the first land snails.

But the Devonian is best known as the "**age of the fish**." Both the cartilaginous fish (e.g. sharks) and the bony fish (e.g. tuna) became abundant. A member of the bony fish, the lobe-finned or **lung fish** known as the Crossopterygians is believed to be the evolutionary source of the first amphibians or the first land vertebrates. The link between the lung fish, the Crossopterygians, and the earliest amphibians is well documented and, therefore, quite convincing. The fossil record shows that their bone structure as well as their tooth structure is quite similar.

The lungfish and the coelacanth, a living fossil thought to be extinct for 70 million years until caught in 1938, both have fleshy, lobed fins that look somewhat like limbs. These two types of fish have long been battling for the honor of which is closer to the ancestral fish that first used fins to walk on land and gave rise to the amphibians which then gave rise to all the other land vertebrates and their descendants (reptiles, birds, and mammals). In order to answer the question as to which fish fist learned to walk on land and breathe air, scientists have recently (in 2013) decoded the genome of the coelacanth. The decoding shows that it's the lungfish, not the coelacanth, which is the closer relative to the first land vertebrates.

The decoding of the coelacanth also showed the presence of a snippet of DNA that enhances the activity of the genes that drive the formation of limbs in the embryo. After inserting the enhancer DNA into mice, an almost normal limb was produced.

It appears, therefore, that we have convincing evidence that the land animals are descended from lobe-finned fish. An enhancer DNA appears to be the mechanism responsible for the development of limbs. Since DNA cannot be reproduced (see Chapter 3(a)) and lies only in the domain of the great designer, we designate the event of the appearance of the first land vertebrates as a **life milestone supportive of the selected seven grand designs**.

(c) The Late Paleozoic Era

The Late Paleozoic in this discussion represents the Mississippian, the Pennsylvanian, and the Permian periods, with time duration of 115 million years.

The Mississippian saw the appearance of a new land plant, the **gymnosperms** (translates into "the naked seed plants"). Unlike the earlier **seedless vascular** plants that had to grow in water only, the gymnosperms, which are also known as the "**flowerless seed plants**," had a reproductive style that freed them from having to stay in or near water. The new plants retained their spores and produced male/female cones with pollen and embryonic seed. This new evolution in land plants is a major life event which eventually opens the door to the development of the "**flowering seed plants**" that dominate the world in all kinds of environments. We designate the appearance of the flowerless seed plants, the gymnosperms, as a **life milestone supportive of the selected seven grand designs**.

The Pennsylvanian period was a time of vast (coal) swamps where conditions were ideal for the **seedless vascular** plants. At the same time, the **gymnosperms or the flowerless seed plants** thrived in non-swamp or dry land areas. Flying insects appeared for the first time, especially thriving in the swampy environments.

During the Pennsylvanian, the amphibians were abundant and became the dominant terrestrial vertebrate animals even though they were required to live near water. The Labyrinthodont amphibians were the most widespread and are believed to have evolved into the first reptiles, the first true terrestrial animals. This evolutionary step, the introduction of a true terrestrial animal that could live in practically any land environment, needed the development of new reproductive process that was independent of water. This process is here referred to as **the miracle of the amniote egg** which allowed the reptiles to completely colonize the land.

Unlike the soft gelatinous eggs laid in water, the amniote egg was protected with a hard shell and could be laid on dry land. In the amniote egg, the embryo is surrounded by the amnion liquid (amnion cavity sac) and is also provided with a yolk sac (food) and a waste sac. This miraculous development of the amniote egg, allowed the reptiles to inhabit all parts of the land because they were no longer required to return to the water for reproduction. This is indeed a grand design event, **Grand Design #5.**

During the Permian, the last period of the Paleozoic, the gymnosperms, the flowerless seed plants, dominated the land. The **seedless vascular** plants declined to a large extent. As for the sea invertebrates, especially the forams, the corals, and the brachiopods, they became once again quite abundant.

At the same time, the reptiles became the dominant land dwellers. Of note here are the *pelycosaurs* (the fin-back or sail-back reptiles), which became quite prominent in the Early Permian, and the *therapsids*, which dominated the landscape during the Middle and Late Permian. The therapsids are considered to have been the most mammal-like of the reptiles. Paleontologists think that therapsids were endothermic (warm blooded) and had a covering of fur.

The dominant reptiles had several advantages for life on land: They possessed an advanced reproduction process, the amniote egg, and advanced jaws and teeth. They were also able to move rapidly on land.

But all this came practically to an end because, at the end of the Permian, the end of the Paleozoic Era, a major extinction event occurred. This extinction event is considered to be **the worst life crisis on earth**. It is estimated that 90% of marine invertebrates, 75% of amphibians, and 80% of reptiles became extinct. Miraculously, the plants were spared. The cause of this major extinction event is still the subject of debate. Probable causes given include the formation of Pangea, and a decrease in salinity. No matter what the cause or causes, the extinction event appears to be part of the overall design, probably the introduction of other animal kingdoms such as the mammals which eventually dominate the earth. We shall refer to the end of the Paleozoic extinction events as yet another **life milestone supportive of the selected seven grand designs**.

Chapter Highlights and Conclusions:

1.   **The Early Paleozoic:**
        The Cambrian Explosion (the sudden appearance of organisms with hard parts) introduced organisms with lots of advantages. The exoskeleton provided protection from predators and radiation but mainly it allowed organisms to grow larger and have advanced locomotion. This explosion of new life is here designated as a life milestone supportive of the selected seven grand designs.

        The appearance of the earliest known vertebrates (armored and jawless fish known as Ostracoderms) represents an evolutionary change that requires the Great Designer. This change has been designated as a life milestone supportive of the selected seven grand designs.

The Taconic Orogeny, a mountain building event relating to plate tectonic activity, caused the extinction of marine invertebrates. This extinction contributed to the great diversification and dominance of vertebrate fish from which eventually evolved the amphibians and the reptiles. This event is designated a life milestone supportive of the selected seven grand designs.

2.   The Middle Paleozoic:

The first appearance of jawed fish (non-marine Acanthodians and marine Placoderms) has been described as "evolutionary opportunism." This event is actually attributed to the Great Designer as a life milestone supportive of the selected seven grand designs.

The earliest non-disputed land plants, the seedless vascular plants, were found in Silurian rocks. The transition from life under water (probably as green algae) to life under the atmosphere (as vascular plants) is difficult to fully understand. The event is here designated as a life milestone supportive of the selected seven grand designs.

The Devonian period is known as the "age of the fish." The lobe-finned lung fish (the Crossopterygians) is believed to be the evolutionary source of the first land vertebrates, the amphibians. The appearance of these first land vertebrates lies in the domain of the Great Designer and is designated as a life milestone supportive of the selected seven grand designs.

3.   The Late Paleozoic:

A new land plant, the *Gymnosperms* (translates into "the naked seed plants") appeared here. The reproductive style of these plants, also known as the "flowerless seed plants," freed them from having to stay in or near water. This evolutionary appearance is designated as a life milestone supportive of the selected seven grand designs.

The *Labyrinthodont* amphibians evolved into the first reptiles, the first true terrestrial animals. This evolutionary change required the development of a new reproductive process that was independent of water. This new process is here referred to as "the miracle of the amniote egg," and is designated as Grand Design #5.

At the end of the Permian period (the end of the Paleozoic Era), a major extinction event occurred. This event is considered to be "the worst life crisis in the earth's history" because it decimated the great majority of the earth's animal inhabitants. This extinction event is designated as a life milestone supportive of the selected seven grand designs.

CHAPTER 6

THE MESOZOIC ERA

The Mesozoic (middle life) represents a time span of 179 million years (see the Geologic Time Scale) during which the reptiles dominated life on earth and, therefore, it is designated as the "Age of the Reptiles." It is also popularly known as the "Age of the Dinosaurs." However, it is also the time during which the birds, the mammals and the flowering plants (the angiosperms) evolved.

The major geologic event of the Mesozoic Era was the break-up of the supercontinent Pangea in the late Triassic period. As noted earlier, Pangea formed at the close of the Paleozoic Era or the end of the Permian period. Thus, the supercontinent Pangea existed as such for several million years before the drifting of the continents resulted in significant separation between them.

The break-up of Pangea mainly affected the oceanic circulation patterns and, therefore, the climate. At the beginning of the Mesozoic, the climate was generally warm and temperate. At the end of the Mesozoic, the climate became   more seasonal and much cooler in general. Also, at the beginning of the Mesozoic, the continental landmass was equally divided above and below the equator. At the end of the Mesozoic, two thirds of the continental landmass moved north or above the equator due to continental drift.

Did all the Pangea break-up effects, outlined above, influence life on earth as well? Yes, as in the case of the Paleozoic, these Mesozoic geologic events affected the evolution of both terrestrial and marine life. Many members of both terrestrial and marine life became victims of the end-Mesozoic mass extinctions which represent the **second worse life crisis in the earth's history.**

For convenience and proper emphasis, the Mesozoic discussion will be subdivided into three sections reflecting the events in terms of the three Mesozoic periods: (a) the Triassic; (b) the Jurassic; and (c) the Cretaceous.

(a)   The Triassic Period

The Triassic period represents time duration of 37 million years.

Some marine invertebrates (e.g. mollusks) and vertebrates (e.g. fish) as well as some land dwellers (e.g. reptiles) who survived the end-Permian mass extinction gave rise to more diverse groups in the Triassic and later periods.

The Cartilaginous fish (e.g. sharks) were abundant but the Bony fish (e.g. tuna) were dominant during the Triassic. Frogs and turtles appeared during this period to replace the devastated numbers of amphibians who perished during the end-Permian mass extinctions. The two main types of land plants (the Seedless Vascular plants and the Naked Seed plants) continued through the entire Triassic but the Naked Seed plants were clearly dominant over the Seedless Vascular. However, the top life story of the Triassic relates to the appearance and the evolution of land vertebrates.

The **Stem Reptiles**, the *Captorhinomorphs* of the Pennsylvanian and Permian periods who were the first to produce the miracle of the amniote egg, first gave rise to the *Thecodonts* (teeth in individual sockets) who, in turn, gave rise to the **Dinosaurs** in Late Triassic. The Thecodonts also gave rise to the relatives of the dinosaurs such as the **Pterosaurs** (flying reptiles), the Ichthyosaurs (the marine reptiles) and other reptiles such as crocodiles, lizards and snakes.

There were two distinct orders of dinosaurs, based on pelvic structure: the **Saurischia** (lizard-hipped) and the **Ornithischia** (bird-hipped). These two orders represent many different species of dinosaurs. Because the dinosaurs became extinct 66 million years ago (end of Mesozoic), people ask the question: Were the dinosaurs poorly adapted animals? Not at all, especially if we consider the fact that the dinosaurs dominated the earth, in all types of environments, for over 140 million years. No other species of land animals comes even close to that. The modern humans have only dominated the earth for about ten thousand years. It's debatable if humans make it to one million or even one hundred thousand years.

Another question often asked about the dinosaurs: Were the dinosaurs lethargic beasts? In other words, were the dinosaurs slow and inactive? A few decades ago, it was believed to be so. However, recent evidence shows some dinosaur species to have been very active and, in fact, some even behaving as if endothermic or warm blooded. Bone structure (dinosaur bones with numerous passageways for blood vessels, typical of endotherms), predator-prey relationships and other features (some caring for their young after hatching) have been cited as evidence for dinosaur endothermy. This characterization of endothermy will support the evidence (presented later in relation to the Jurassic period) that birds were derived from dinosaurs.

As pointed out above, the **stem reptiles** of the Paleozoic Pennsylvanian period were the ancestors of the dinosaurs and their relatives. These same stem reptiles also gave rise to the primitive mammal-like reptiles of the late Paleozoic (the Therapsids) which, in turn, gave rise to the advanced mammal-like reptiles (the Cynodonts) which gave rise to the **mammals**.

The earliest mammals evolved during the late Triassic but they are difficult to distinguish from their reptilian ancestors, the Cynodonts. This close relationship between the advanced reptiles and the early mammals is referred to as **mosaic evolution** which implies that the mammals had both reptilian and mammalian characteristics. For instance, the early mammals, the Monotremes, reproduced through egg laying. The diversity of the Mesozoic mammals was low and they were generally the size of rats, with a few the size of a raccoon. As we shall see below, the more advanced mammals, the Marsupials (use of the pouch) and the Placental (use of the placenta) evolved during the Cretaceous period.

The appearance of the dinosaurs and their relatives (the pterosaurs and the ichthyosaurs and other reptiles) and the appearance of the early mammals, both are attributed to evolution. As pointed out in the Background Information on evolution, scientists outlined the **modern synthesis of what is evolution**: natural selection is a mechanism that accounts for evolution, the chromosome/gene theory is a mechanism that causes evolution, and the variations of sexual reproduction and mutation are the two main causes that initiate evolution.

The **modern synthesis**, outlined above, is really a compilation of suggestions or ideas or theories of how species evolve. The synthesis cannot explain where the original parts (the cells, the DNA, the genes) come from and how exactly the required complex chemical or biological mechanisms originate or work. These evolutionary changes require a master design. We, therefore, classify these evolutionary changes as a **life milestone supportive of the selected seven grand designs.**

(b)   The Jurassic Period

The Jurassic period represents time duration of 64 million years.

During this period, the crocodiles appear and become the dominant amphibians. No major changes in fish (the bony fish still dominant) or in land plants (the gymnosperms or naked seed plants still dominant). The most important life developments appeared in the category of land vertebrates.

The dinosaurs became most abundant and diverse during this period and also during the next (the Cretaceous). Both orders of the dinosaurs, the Ornithischia or bird-hipped and the Saurischia or lizard-hipped, dominated the land. The Ornithischia were all herbivorous and ranged in length from less than a meter to 9 meters. The best known and most widespread were the genus Anatosaurus or the Duck-bills who, unlike typical reptiles, used colonial nesting and cared for their young. The Saurischia included both carnivorous and herbivorous and ranged in length from less than a meter to over 28 meters. The genus Tyrannosaurus Rex or T-Rex is believed to have been one of the largest terrestrial carnivores that ever existed. The herbivorous genus Brachiosaurus represents the largest land animal that ever existed.

Also abundant during the Jurassic period (and the Cretaceous) were the relatives of the dinosaurs, the flying reptiles (the Pterosaurs) and the marine reptiles (the Ichthyosaurs and the Plesiosaurs).

The Pterosaurs were the first vertebrate animals to fly. The early species were sparrow size but later species, in the Cretaceous, had a 12 meter wing span. Wing membranes, supported by a 4th finger, as well as light hollow bones and better brains, for muscular coordination and sight, were among the developments that aided in their flight adaptations. Experimental studies indicate that the smaller Pterosaurs were wing-flapping fliers. The larger ones could fly but once airborne they would glide. At least one small Pterosaur and other dinosaurs were found to have a coat of hair or hair-like feathers. This indicates that they were probably endothermic (warm-blooded). This fact also provides evidence of the relationship between dinosaurs and birds.

The other dinosaur relatives, the marine reptiles, also thrived during the Triassic period and the rest of the Mesozoic. The fish-eating Ichthyosaurs (fish reptiles) were completely aquatic and looked like the present-day porpoises. Because their reptile eggs would not survive if laid in water, the females retained the eggs and gave birth to live young. Fossils were found with young ones in the belly. The Plesiosaurs, on the other hand, looked like the present-day walruses and seals. They had oar-like limbs, and laid their eggs ashore.

Birds appeared during the Mesozoic. The oldest known bird, *Archaeopteryx* (ancient feather) was found in Jurassic rocks of Germany. It appears that this first bird evolved from small carnivorous dinosaurs of the Saurischia (lizard-hipped) order, not the Ornithischia (bird-hipped) order as one may surmise. This first bird is an excellent example of **mosaic evolution** in that it retains a mosaic of characteristics from its ancestors, the dinosaurs, as well as a mosaic of bird characteristics. More specifically, *Archaeopteryx* has dinosaur-like teeth, tail, hind limbs, and brain size but it also possesses feathers and a wishbone. The teeth in birds and other dinosaur-like features were completely lost by the end of the Mesozoic. Birds represent a major category of life on earth. Their first appearance in the Jurassic period is hereby denoted as another **life milestone supportive of the selected seven grand designs**.

(c) The Cretaceous Period

The Cretaceous period represents time duration of 78 million years.

During this period, the dinosaurs and their relatives (flying reptiles and marine reptiles) obtained their greatest diversity and dominance over the land. However, at the end of the period, the dinosaurs and their relatives as well as several groups of marine invertebrates became the victims of mass extinctions. These end-Mesozoic extinctions represent the **second worst life crisis** on earth. As pointed out earlier, the worst life crisis was at the end of the Paleozoic, the end of the Permian period.

It appears that extinctions are the rule of life. Scientists believe that over 99% all species that ever lived are now extinct. Some probable causes of extinctions are disease/epidemics, climate changes, food shortages, changes in the composition of the atmosphere, and, in our time, extinctions brought about by humankind. Humans have a degree of control, over the environment, which is unprecedented in the history of life.

The meteorite impact hypothesis has been the most popular single cause usually associated with the extinction of the dinosaurs. The meteorite impact scenario goes something like this: (a) A meteorite of at least 10 kilometers in diameter impacted the earth. (b) Dust, about 60 times the mass of the meteorite, was blasted high into the atmosphere. (c) Heat, generated at impact, caused fires which sent more matter (ashes) into space. (d) Sunlight was blocked for several months and the earth's surface temperature was reduced. (e) The absence of sunlight caused the temporary cessation of photosynthesis and the collapse of food chains. (f) Extinctions followed.

There is physical evidence to support a large meteorite impact at the end of the Cretaceous. For instance, thin clay layers at the borderline between the Cretaceous and the Tertiary periods are high in Iridium (Ir^{77}), an isotope high in meteorites and rare in crustal rocks. Also, soot and shock-metamorphosed quartz grains are found at some of these borderlines, suggesting an impact.

There are of course problems with the meteorite impact hypothesis if one assumes that the impact was the only cause of the extinctions. The main problem is the selective nature of the extinctions. For instance, why would such an impact not affect the crocodiles (relatives of the dinosaurs) or even tropical plants which are sensitive to cold temperatures? Both the crocodiles and the tropical plants were spared from extinction.

Because of the latter and other reasons, many paleontologists think that other factors were also important. For instance, climates became harsher at the end of the Mesozoic. Also, paleontologists believe that the dinosaurs were already on the decline towards the end of the Mesozoic and headed for extinction. The meteorite impact is believed to have hastened the process of extinction.

Whatever the cause or causes of the extinction of the dinosaurs, the life crisis resulting from the process provides an opportunity for a new class of animals, the mammals, to dominate life on earth. It appears that every change in life is part of the grand design.

The earliest mammals, the Nonotremes which laid eggs, evolved in late Triassic (see above). More advanced mammals, the Marsupials (the pouches) and the Placental evolved during the Cretaceous. The placental became the dominant life on earth during the Cenozoic, the new and latest era of life on earth. Why did the placental eventually dominate the earth (90% of all animals)? Primarily because of a miracle that happened during the Cretaceous: The introduction of the **placenta**, a new reproductive method where the young develop fully before birth. The placenta is here considered as a miraculous event, **Grand Design #6**.

Chapter Highlights and Conclusions:

1.   **The Triassic:**

The dinosaurs and their relatives, the flying reptiles (Pterosaurs) and the marine reptiles (Ichthyosaurs) and other reptiles (crocodiles, lizards and snakes), all evolved during the Late Triassic. Many different species of dinosaurs, representing the Saurischia (lizard-hipped) and Ornithischia (bird-hipped) orders, dominated the earth, in all types of environment, for over 140 million years, by far more than any other species that lived on earth. Some species are believed to have been endothermic or warm blooded.

The earliest mammals also evolved during the Late Triassic. Their ancestors, the Stem Reptiles of the Late Paleozoic (Pennsylvanian) were also the ancestors of the dinosaurs and their relatives. The appearance of the early mammals and that of the dinosaurs and their relatives, both are attributed to evolution.

The modern synthesis of what evolution is (see p. 43) cannot fully explain where the required parts (the cells, the DNA, the genes) come from and how exactly the complex chemical and biological mechanisms originate or work. These evolutionary changes require a master design. We, therefore, classify these evolutionary changes as a life milestone supportive of the selected seven grand designs.

2.   The Jurassic:

The dinosaurs and their relatives, the flying reptiles (the Pterosaurs) and the marine reptiles (the Ichthyosaurs and the Plesiosaurs) became most abundant and diverse. In this period, most of the early species of the Pterosaurs, the first vertebrate animals to fly, were quite small (sparrow size).

The oldest known bird, Archaeopteryx (ancient feather) was found in Jurassic rocks of Germany. Evidence shows that this first bird evolved from small carnivorous lizard-hipped dinosaurs and is an excellent example of mosaic evolution: It retains characteristics from its ancestors, the dinosaurs, such as teeth, tail, hind legs, and brain size but it also possesses feathers and a wishbone.

Birds represent a major category of life on earth. This first appearance in the Jurassic is denoted as a life milestone supportive of the selected seven grand designs.

3.   The Cretaceous:

During this period, the dinosaurs and their relatives (flying reptiles and marine reptiles) obtained their greatest diversity and dominance over the land. However, at the end of this period, the dinosaurs and their relatives, plus several groups of marine invertebrates, became the victims of mass extinctions which represent the second worst life crisis on earth.
The Meteorite Impact Hypothesis has been the most popular cause usually associated with the extinction of the dinosaurs at the end of the Cretaceous period (end of the Mesozoic Era). In fact, there is physical evidence to support a large meteorite impact at that time. However, paleontologists believe that other factors, such as climate changes, were also important as causes of the mass extinctions.

More advanced mammals, the Marsupials and the Placental, evolved during the Cretaceous. With the extinction of the dinosaurs and their relatives at the end of the Mesozoic, the Placental became the dominant life on earth during the next and last era, the Cenozoic. The great advantage of the Placental was their new reproductive method, the placenta, where the young develop fully before birth. The introduction of the placenta is considered a miraculous event in the history of life, and is designated here as Grand Design #6.

CHAPTER 7

THE CENOZOIC ERA

The Cenozoic (new life) represents a time span of 66 million years (see the Geologic Time Scale, p. 71) during which the continued rifting of supercontinent Pangea produced the present distribution of continents and oceans. Although the Cenozoic time span is relatively brief, it includes not only significant evolution of the earth itself but also significant evolution of the earth's biosphere or life on earth.

For convenience and proper emphasis, the Cenozoic discussion will be subdivided into two sections reflecting the geologic and biologic events in terms of the two Cenozoic periods: (a) the Tertiary; and (b) the Quaternary.

(a)   The Tertiary Period

The Tertiary period represents time duration of a little over 64 million years.

World geologic events, resulting from the continued rifting of Pangea, include: 1) The Alpine Orogenic (mountain building) Belt and 2) The Himalayan Orogenic Belt.

During the Alpine Orogenic Belt, the African Plate collided with the Eurasian Plate (the European section of the plate) to form the Alps and other European mountains (e.g. the Atlas of N. Africa, the Balkan and other Eastern Mediterranean mountains). During the Himalayan Orogenic Belt, India collided with Eurasia (the Asian section of the plate) to form the Himalayan Mountains.

North American geologic events involve primarily the North American Cordilleran Mountains (from Alaska to Mexico) in terms of the deformation they suffer during the Laramide Orogeny, mostly affecting the Rocky Mountains. Also, Cordilleran volcanism continued throughout the Cenozoic, culminating in the formation of the Columbian River basalts, one of the world's greatest eruptive events, long fissure flows covering the Northwest with basalt up to 4,000 feet thick.

In terms of life, the Tertiary period is known as the **Age of Mammals**. The history of mammals is better known than any other class of vertebrates because the Cenozoic terrestrial deposits are more common and more accessible (last to be deposited on the surface), and the mammalian fossils are easier to identify primarily because the mammalian teeth are fully differentiated into distinctive types (e.g. chewing molars different in each order of mammals).

It was noted above, that the placental mammals evolved during the Cretaceous period of the Mesozoic Era, and the introduction of the placenta at that time was designated as Grand Design #6. The Cretaceous placental mammals were the ancestors for the placental mammalian orders that evolved during the Cenozoic.

About 90% of all animals that live on earth are placental mammals. Their success relates to their reproductive method, the miracle of the placenta, which allows the offspring to develop before birth.

The evolutionary history of one placental mammal, the horse, is exceptionally well documented by fossils found in superposed layers representing most of the Cenozoic. The primary evolutionary change in horses relates to their size. The earliest horse was the size of a fox before it acquired its present size. Other adaptations among hoofed mammals include teeth modifications for grinding vegetation and limb modifications for speed.

Perhaps the most important life development during this period was the appearance of the order of the **Primates** from which eventually evolved the family of the **Hominids** which includes the **human relatives** and the **humans**. Several unique characteristics differentiate the Primates from other mammalian orders. These characteristics include their skeletal structure and mode of locomotion, a larger brain, stereoscopic vision, and grasping hands with an opposable thumb.

The flowering plants (the angiosperms) continued to dominate the land plant communities throughout the Tertiary and the rest of the Cenozoic. Fully modern birds (no teeth) appeared during early Tertiary. They increased in diversity and acquired numerous habitats throughout the Cenozoic.

All these major evolutionary changes, which led to the eventual dominance of the modern humans, are hereby designated as a **life milestone supportive of the selected seven grand designs.**

(b)   The Quaternary Period

The Quaternary period represents time duration of 1.8 million years (see the Geologic Time Scale, p. 71). It is approximately the last half minute of a 24-hour day of the earth's history. Time-wise it appears relatively insignificant and yet this is the period during which our own species, the *Homo Sapiens*, evolved.

During the Quaternary period, several major glacial and interglacial episodes, part of the period's Ice Ages, generally known as "Pleistocene Glaciation," occurred throughout the world. Large ice sheets advanced and retreated on most of the earth's surface, significantly affecting the earth's present-day topography (the shape of the earth's surface). At the same time, some of the human relatives and, eventually, the humans evolved.

Based on terrestrial glacial deposits and erosional features created by the glaciers, scientists determined that there were 4 major glacial-interglacial episodes in North America and 7 such episodes in Europe. Several warm-cold Pleistocene climatic cycles are recognized through paleontologic data (e.g. pollen analysis) or oxygen isotope O^{16}/O^{18} data (e.g. deep-sea core analysis). By plotting "time" versus "temperature" variations during these cycles, it has been determined that the main factor that separates the glacial and interglacial episodes noted above (4 episodes in North America and 7 in Europe) is a difference of 5°C world average temperature: 5 degrees cooler during a glacial cycle or 5 degrees warmer during an interglacial cycle.

There is no doubt that climatic changes are the main causes for the Pleistocene Glaciation. However, there is no simple answer as to what causes these climatic changes. We can outline some probable causes for these climatic changes but it should be noted that different mechanisms assumed responsible for climatic changes interact between them in complex ways. Here is an outline of the 3 most probable contributing causes of climatic change that may contribute to one extent or another towards a glacial episode:

1. **Volcanic eruptions**. These are considered short-duration events. Their contribution to climate change is relatively small compared to the other two outlined causes listed below. Volcanoes may produce great quantities of dust and gases that enter the atmosphere and block the sun's energy from reaching the earth's surface, thus reducing the earth's atmospheric temperature. Historical examples of such volcanic eruptions include the 1815 Mt. Tamboa (Indonesia) eruption that caused the coldest summer in the Northern Hemisphere (snow in the summer) and the 1883 Krakatoa (Indonesia) eruption that caused the world average temperature to drop by the significant amount of 2°C.

2. **Plate Tectonic Activity**. Such activity may produce long-duration effects. The opening and closing of oceans and especially the movement of continents to higher latitudes or the creation of mountain chains or higher altitude through collision or subduction (see Plate Tectonics in the Background Information section, p. 79). Thus, higher latitude or altitude may provide the right environment of the accumulation of snow and the beginning of glacier formation.

3. **Astronomical Causes (the Milankovitch theory).**
During the 1920's, Milankovitch, a Serbian astronomer,
proposed that minor irregularities in the Earth's rotation
(e.g. tilt of axis) and orbit around the Sun (e.g. geometric
shape of the orbit) alter the amount of solar radiation
received at any given latitude and, therefore, cause
intermediate-term climatic changes. Thus, continuous
variations in the Earth's orbit and axial tilt are now generally
believed (by most scientists) to cause complex climatic
changes and most probably provided the triggering
mechanism for the glacial – interglacial episodes of the
Pleistocene epoch.

The ice ages changed a large portion of the earth's
topography significantly and affected the distribution and
survival of life on earth. In a way, these glacial events
prepared the stage for the eventual emergence of the
modern humans. These Quaternary events are, therefore,
part of the great design and are designated as a **life
milestone supportive of the selected seven grand designs.**

As in the case of the Tertiary period, **life** in the
Quaternary period is well documented primarily because it
involves the most recent and relatively shortest period in the
earth's history. The fossils are usually better preserved and
more accessible because of their proximity to the surface.
Our discussion of Quaternary life will be devoted to human
relatives and humans, an evolution that began with the
appearance of the mammalian order of the Primates in early
Cenozoic.

The **Primates order** is divided into two suborders: The **Prosimians suborder** (lemurs, tarsiers, tree shrews) and the **Anthropoids suborder** (the Old and New World monkeys, apes, gibbons and siamangs, and human relatives and humans). We shall concentrate on the Anthropoids suborder which is divided into three superfamilies, one of which is the **Hominoids superfamily**. From this superfamily evolved the **Apes Family** (chimps, gorillas, orangutans) and the **Hominids Family** (human relatives and humans). This shows that the apes and the humans had similar ancestors but they are NOT directly derived from one another as some suggested at various times.

The potential ancestors of modern apes and humans (derived from the **Hominoid superfamily**) cover a range from 20 million years ago to 7 million years ago. The fossil record of the **Hominids family** (human relatives and humans) extends back to only about 4 million years ago. The interval, 7 – 4 = 3 million years, is the **missing fossil link** or the time when the **Hominids** (humans) diverged from the **Hominoid** ancestors (apes and human relatives). For decades scientists have been searching for the missing fossil link but none of the findings is generally accepted to represent the elusive missing link.

It appears that the missing fossil link is just another indication of the uniqueness of the human family and its introduction as part of a design by the great designer. It is here designated as a **life milestone supportive of the selected seven grand designs**.

Finally, a few words about the **Hominids family** , the family of the human relatives and humans. Some of the more distinguishing characteristics of the family members included their upright posture, their large and complex brain, their reduced face and canines, and their manual dexterity (ability to make and use tools). Anthropologists divide the Hominids family into the **genus Australopithecus** and the **genus Homo.**

The genus Australopithecus, a fully bipedal group that evolved in Africa, is divided into five species which range in age from 4.2 to 1.2 million years ago. One of the most complete skeletons of human relatives ever found includes that of a young girl (named Lucy by anthropologists) belonging to the Australopithecus species Afarensis.

The genus Homo is divided into three species: The species *Homo Habilis, Homo Erectus*, and our own species, the *Homo Sapiens*. The ape-like *Homo Habilis* lived in Africa from about 2.5 to 1.6 million years ago. *Homo Erectus* evolved from *Homo Habilis* in Africa but was the first Hominid to migrate out of Africa into Europe and Asia. *Homo Erectus* had an upright posture and a massive face, lived in caves, and used tools and fire. His range extended to 0.25 million years ago. It is unclear whether the transition from *Homo Erectus* to *Homo Sapiens* took place in Africa or in a multi-region outside Africa. What is clear is that, by this time when the humans or *Homo Sapiens* appeared, all the previous human relatives became extinct.

The first member of our *Homo Sapiens* species is the
Neanderthals who inhabited Europe and the Near East
between 200,000 and 30,000 years ago. They lived in hut-like
shelters and caves and made specialized tools and weapons.
Fossil findings suggest that they buried their dead. In
comparison to the modern humans in terms of appearance,
the Neanderthals were more robust and hairy and had
differently shaped skulls.

The second member of our *Homo Sapiens* species is
the **Cro-Magnons**. They lived from about 32,000 to 10,000
years ago and are considered the successors of the
Neanderthals. In terms of appearance, they resembled
modern humans to a great extent. They were highly skilled
nomadic hunters, they formed living groups of various sizes,
and they had art (cave painters) and technology. As pointed
out above, they are thought to have replaced the
Neanderthals but it is not at all clear exactly when and
especially how. Based on their range of existence, the Cro-
Magnons co-existed with the Neanderthals. Why the
Neanderthals became extinct is a big mystery.

The third and last member of our *Homo Sapiens*
species is the **Modern Humans**. They succeeded the Cro-
Magnons about 10,000 years ago. They have since spread out
throughout the world and even travelled into outer space,
including landing on the Moon. They invented writing
about 5,000 years ago and, therefore, they now have
recorded history. They also have achieved incredible artistic,
scientific, technological, and medical advances.

The appearance and exceptional development of the Modern Humans, during the last few seconds of a 24 - hour world existence, represent a continuous and miraculous design by the Great Designer. The age of the Modern Humans is, therefore, designated as Grand Design#7.

Two final comments on human evolution: (1). In1998, scientists conducted DNA tests on the original Neanderthal fossils found in Germany. The conclusion, although it may not be generally accepted, is that the Neanderthals represent a parallel evolutionary development and are not directly related to Cro-Magnons and Modern Humans. (2). Molecular anthropologists conducted a search for the elusive Eve, everybody's mother, and found one who lived in the hot savannas of Africa, 200,000 years ago. She left resilient genes carried by all humans, Mitochondrial DNA, which are inherited only from the mother and, therefore, are useful for tracing family trees and preserving the family record. Recently, similar fossils were found in other parts of the world.

Chapter Highlights and Conclusions:

1.   **The Tertiary:**

This period is known as the <u>Age of the Mammals</u>. The placental mammals actually evolved during the previous period (the Cretaceous of the Mesozoic Era) but became dominant during the Cenozoic Era. About 90% of all animals that live on earth today are placental mammals. Their success relates to their reproductive method, the placenta, whose introduction was designated as Grand Design # 6 (see Chapter 6 (c), the Cretaceous Period).

A most important life development during this period was the appearance of the order of the <u>Primates</u> from which evolved the family of the Hominids which includes <u>the human relatives and the humans</u>. These evolutionary developments plus others (e.g. fully modern birds) are designated as a life milestone supportive of the selected seven grand designs.

2.   The Quaternary:

During this period, several major glacial and interglacial episodes, representing the period's Ice Ages, occurred throughout the world. Based on terrestrial glacial deposits and erosional features created by the glaciers, it was determined that there were 4 major glacial-interglacial episodes in North America and 7 such episodes in Europe.

Climatic changes are believed to be the main causes for these Ice Ages. Probable causes of climatic change that may contribute towards a glacial episode include volcanic eruptions (dust and gases enter the atmosphere and block the sun's energy), plate tectonic activity (movement of continents to higher latitudes or the creation of mountain chains), and astronomical causes such as minor changes in the earth's rotation (tilt of axis) or changes in the geometric shape of the earth's orbit around the Sun.

The Ice Ages changed a large portion of the earth's topography significantly, and affected the distribution and survival of life on earth. They are part of the great design and are designated as a life milestone supportive of the selected seven grand designs.

The Primates order (evolved in the previous period, the Tertiary) is divided into two suborders, one of which is the <u>Anthropoids suborder</u> which includes monkeys, apes, and human relatives and humans. The Anthropoids suborder is divided into three superfamilies, one of which is the <u>Hominoids superfamily</u>. From this superfamily evolved the Apes Family (chimps, gorillas, orangutans) and the <u>Hominids Family</u> (human relatives and humans).

The Hominoids superfamily covers a range from 20 million to 7 million years ago but the Hominids family extends back to only 4 million years ago. The <u>missing fossil link of 3 million years (7-4=3)</u> is the time when the Hominids (humans) diverged from the Hominoid ancestors (apes and human relatives). The missing fossil link is an indication of the uniqueness of the human family whose introduction by the great designer is hereby designated as a life milestone supportive of the selected seven grand designs.

Anthropologists divide the Hominids family into the genus Australopithecus and the <u>genus Homo</u>. The five species of the genus Australopithecus, a fully bipedal group that evolved in Africa, disappeared around 1.2 million years ago. The <u>genus Homo</u> is divided into 3 species: The Homo Habilis (ape-like), the Homo Erectus (upright posture and massive face), and our own species, the <u>*Homo Sapiens*</u>. By the time the humans (Homo Sapiens) appeared (around 200,000 years ago), all human relatives (all the species of the Homo Habilis and the Homo Erectus) became extinct.

The <u>Neanderthals</u> represent the <u>first member</u> of our *Homo Sapiens* species. They inhabited Europe and the Middle East between 200,000 and 30,000 years ago. They lived mostly in caves and made specialized tools and weapons. In comparison to modern humans, they were more robust and hairy and had differently shaped skulls.

The <u>Cro-Magnons</u> represent the <u>second member</u> of our species. They lived from about 32,000 to 10,000 years ago, resembled modern humans to a great extent, and are thought to have replaced the Neanderthals with whom they coexisted for a couple of thousand years. It is not at all clear as to exactly when and especially how they replaced the Neanderthals. Also, why the Neanderthals became extinct is a very big mystery.

<u>Modern Humans</u> represent <u>the third and last member</u> of our species. They succeeded the Cro-Magnons about 10,000 years ago. They have since spread out throughout the world and even travelled into outer space. They achieved incredible artistic, scientific, technological, and medical advances.

The appearance and exceptional development of the Modern Humans, during the last few seconds of a 24 – hour world existence, represent a continuous and miraculous design by the Great Designer. The age of the Modern Humans is, therefore, designated as <u>Grand Design#7</u>.

BACKGROUND INFORMATION

Minerals and Rocks

A **mineral** is a naturally occurring native element or atom (e.g. Au or gold) or compound (e.g. SiO_2 or quartz) that forms from inorganic processes (as opposed to organic or life processes). Its definite chemical composition and the geometric arrangement of its atoms reflect certain environmental conditions (e.g. temperature or pressure) that effect its formation or destruction. On the basis of the latter definition, mineralogists have identified close to 4,000 minerals on earth. Only 10 of these minerals are truly common, and 7 out of these are silicates (contain Si and O in their composition such as quartz, SiO_2, or feldspar, $KA_lSi_3O_8$). The rest are carbonates such as calcite ($CaCO_3$), halides such as halite or "salt" (NaCl) and sulfates such as gypsum ($CaSO_4$).

A **rock** is usually made up of two or more minerals (e.g. granite, with two essential minerals – quartz, SiO_2 and K-feldspar, $KA_lSi_3O_8$) or it is monomineralic (e.g. marble, only calcite, $CaCO_3$). Regolith represents the loose materials on the surface of the earth (usually referred to as soils), and bedrock represents the continuous solid rock that underlies regolith. Bedrock is divided into 3 major rock groups or major rock types based on two factors: 1. The process that forms these rocks, and 2. Whether these rocks form on the surface of the earth or at depth.

Igneous rocks (from the Latin for "fire") form through the process of **cooling of a melt**. Molten rock or a melt results from the melting of the solid bedrock, usually at considerable depth below the surface. The melt starts cooling as it moves towards the surface (the area of least pressure). If the melt reaches the surface, it cools faster, and is referred to as **lava**. If the melt does not reach the surface, it cools slowly, and is referred to as **magma**.

Fast cooling lava forms the **fine-grained igneous volcanic** rocks such as rhyolite or basalt on the **surface**, whereas, slow cooling magma forms the **coarse-grained igneous plutonic** rocks such as granite or gabbro **at depth**. Volcanic igneous rocks are also known as **extrusive**, while plutonic igneous rocks are also known as **intrusive**.

Sedimentary rocks (from the Latin for settling) form through the process of **diagenesis** which essentially involves the accumulation, compaction, and cementation of sediments or particles on the **surface only**, Typically this surface represents a water basin (bottom of an ocean or a lake or a river) or a land surface. Sedimentary rocks may also form as **chemical or biochemical precipitates**, usually at the bottom of the ocean.

Metamorphic rocks (from the Greek for change-form) relate to a process of formation that involves heat, directional or differential pressure, and chemical fluids (each to one degree or another) acting on pre-existing rocks found at great depth. The process causes changes of composition or texture to pre-existing rocks in the solid state (without melting them).

Fossils and Relative Geologic Age:

A **fossil** represents either the body remains (usually hard parts or bones) or the traces of the activity (such as footsteps or nests) of a once-living organism. A fossil implies antiquity, usually at least a thousand years. Fossils are mostly found in layered **sedimentary rocks** that usually form at the bottom of water basins. Fossils do not normally survive in igneous or metamorphic rocks because of immense heat or pressure (see Introduction on rocks above). Thus, fossils not only provide evidence of life in the earth's past but they also help us determine past depositional environments. The sedimentary layers may also provide information on **relative geologic age** (i.e. that one layer is simply younger or older than another layer without knowing their numerical age) through the use of certain geologic principles such as the *Principle of Superposition* which tells us that in a sequence of undisturbed sedimentary layers, the layer at the bottom of the sequence is the oldest and the layer on top of the sequence is the youngest.

The Geologic Time Scale:

Through the use of fossils, which show changes in organisms through time, and geologic principles, which provide information on relative age, many geologists have examined a large number of rock formations around the earth and, thereby, created the **Geologic Time Scale** (see p. 71). The Geologic Time Scale lists rock formations on the basis of relative geologic age, always displaying the older rock formations at the bottom of the scale and the younger formations at the top of the scale. The periods in the Geologic Time Scale usually have geographic names of areas (around the world) whose specific fossils and other rock characteristics have placed them in a certain position on the time scale (see a modified reproduction of the Geologic Time Scale at the end of this section). For instance, the **Cambrian** represents the oldest period of the **Paleozoic Era** ("ancient life") because it was first discovered in the Cambrian area of Wales. The **Mississippian** and **Pennsylvanian** periods represent areas in the two American states. Thus, we can **correlate or establish age equivalence through similarity of fossils** by, for instance, describing a rock formation in Australia as being of Cambrian or Mississippian age.

When first invented, the Geologic Time Scale provided information on relative time only. With the 1895 discovery of **radioactivity** (the spontaneous decay of any radioactive or unstable atomic isotope by emitting particles of itself ,and heat, and eventually resulting in the production of a new and stable isotope), geologists have been able to assign **absolute time** or numerical ages to the subdivisions of the Geologic Time Scale, usually denoted as millions of years BP (Before the Present).

The principle in determining absolute time is relatively simple once we make certain assumptions about the rate of radioactive decay (assuming it to be constant through time) and other factors, especially if we assume a near closed system (essentially correcting for or minimizing the effects of processes such as weathering and erosion) when we determine (through the use of a spectroscope) the amount of a certain isotope in a mineral or rock. The basic formula for determining isotope amounts is: $P_0 = P_t + D$, where P_0 represents the amount of the radioactive parent isotope at time zero or when it starts out, Pt represents the amount of the parent isotope that did not decay or was left over in the mineral at a later time, say today, and D represents the amount of the new stable or daughter isotope that was formed during the decay process.

The numerical ages listed for the eras and periods as well as the Precambrian eons (see the Geologic Time Scale, next page) represent some of the most generally accepted isotopic determinations published. The beginning of the time scale (4,600 my BP) represents the age of the earth, primarily obtained through the isotopic dating of rocks in the earth's crust as well as moon rocks and meteorites and through the use of astronomical data such as the "red shift."

THE GEOLOGIC TIME SCALE[1]

ERA	PERIOD	Mill.YEARS[2]	MAJOR EVENTS[3]
CENOZOIC		0.01	Modern humans; ice age ends
(new life)	Quaternary	1.8	Stone-age humans; ice age begins
		3	Earliest humans
		24	Flowering plants (angiosperms) dominant
		37	Apes appear
		58	Horses appear; formation of Alps & Himalayas
	Tertiary	66	Mammals and birds abundant
MESOZOIC	Cretaceous	144	Dinosaur extinction (Second worst life crisis)
(middle life)			Flowering plants (angiosperms) appear
	Jurassic	208	First birds & mammals (the placenta)
	Triassic	245	Dinosaurs appear; the break-up of Pangea
PALEOZOIC	Permian	286	Extinctions (worst life crisis); Pangea forms
(ancient life)	Pennsylvanian	320	Reptiles evolve (the amniote egg)
	Mississippian	360	Naked seed plants (gymnosperms) appear
	Devonian	408	Age of fish; amphibians appear
	Silurian	438	Earliest land plants (seedless vascular) appear
	Ordovician	505	Diversification of invertebrates and fish
	Cambrian	570	First shelled invertebrates; vertebrates (fish)
PRECAMBRIAN	Proterozoic Eon	2500	Earliest fossils of the eukaryotic cell
	Archean Eon	4600[4]	Earliest fossils of life on earth (prokaryotic)

1 Condensed and modified or simplified for this discussion
2 Millions of years BP (before the present). From the Geological Society of America for the Decade of North American Geology (1983)
3 Selected major earth and life history events
4 The age of the earth

Evolution:

Evolution is biological change, through time, that is heritable (i.e. it passes from parent to offspring). A scientific *theory* represents scientific explanations of natural phenomena with lots of supporting evidence. The **general theory of evolution** is based primarily on evidence from fossils such as the fact that older and older fossiliferous layers or sedimentary rocks contain fossils of organisms increasingly different from those in younger and younger rocks or those living today. Other evidence includes comparative anatomical structures such as vestigial structures (the questionable present-day use of a dewclaw on the forefoot of a dog, or an appendix in humans) or observations of small-scale evolution (one that can be observed in a lifetime) of modern organisms such as rodents or insects developing resistance to pesticides and insecticides. The theory is generally acceptable and it states that all living organisms are the evolutionary descendants of life forms that existed in the past.

A specific theory of evolution such as **divergent evolution**, which claims that all living organisms descended from a common ancestor, may be controversial and not widely acceptable. Here we endorse the general theory but we shall also refer to certain specific meritorious theories that account for evolution (e.g. Darwin/Wallace) or processes or mechanisms that cause evolution (e.g. Mendel), all representing significant published contributions of the last two centuries.

Well into the 18th century, the prevailing belief was that all important knowledge was contained in the works of Aristotle and the Bible. A popular concept, the *fixity of species*, proclaimed that all species had been created in their present form and had not changed through time. Things began to change, towards the end of the century, with the appearance of the **Enlightenment** philosophical movement that relied on rationalism rather than divine revelation. Geologic principles and the great age of the earth were becoming acceptable. Evolution, as a biological change through time, from one species to another, became an acceptable idea. (It is interesting to note that the idea of evolution was never intended to explain how life originated, only how life has changed through time.) However, a theory to explain evolution was lacking.

The first popular theory was that of **Lamarck**, a botanist/geologist, in the first decade of the 19th century. The theory was described as the *inheritance of acquired characteristics* which states that organisms acquire traits or characteristics during their lifetime and then pass on these acquired characteristics to their offspring. Lamarck explained the capacity of giraffes to produce offspring with long necks as a trait acquired through the habit of neck stretching (short-necked giraffes stretched their necks in order to reach leaves on tall trees). The theory appeared "logical" to most people, thus it was widely accepted. Today we know that all organisms possess hereditary determinants (genes) that cannot be modified by any effort of an organism during its lifetime.

Other theorists, such as Cuvier, followed. He was a critic of Lamarck and believed that each different fossil in each different sedimentary layer represented a separate creation. However, the two that made an impact with a theory that is still in effect or generally acceptable today are **Charles Darwin** and **Alfred Wallace**. The two workers came up with a similar idea, independently obtained and simultaneously presented in 1859. However, Darwin's publication **Origin of Species** (1859) and his theory called **Natural Selection**, sometimes referred to as "the survival of the fittest," are better known.

Darwin presented his theory as a mechanism that accounts for evolution with four critical points: 1. All organisms possess heritable variations such as size, shape, color, etc. 2. Some variations are more favorable or provide an edge. 3. Those organisms with favorable variations, or an edge, are more likely to survive. 4. Those who survive pass on their favorable characteristics to their offspring.

The giraffe's " long neck" according to natural selection by Darwin/Wallace may be explained as follows: Those giraffes with long necks have a favorable characteristic or an edge in terms of reaching more leaves, and they are, therefore, more likely to survive and reproduce, passing the long neck variation to their offspring.

Darwin documented evidence for evolution but had no process or mechanism **to cause** evolution to occur. He leaned toward Lamarck's process but also recognized its weakness. Finally, in 1866, Gregor Mendel, an obscure Austrian monk, published his experimental work with garden peas of several variations (color, shape, etc.) and introduced a mechanism that causes evolution to occur, using math to explain his breeding observations. Mendel sent a copy of his work to Darwin who never read it. Mendel's work lay buried until it was discovered in 1900. Essentially, Mendel discovered that offspring occur in fixed ratios of their parents' variations and also that characteristics or traits are passed to offspring by discrete units now called **genes**. More specifically, today we know that the cells of all organisms (except bacteria) contain thread-like structures called chromosomes which are molecules of DNA. Specific segments of the DNA molecule are the basic hereditary units, the genes.

In the 1940's, paleontologists, geneticists and population biologists outlined the **modern synthesis** of what is evolution: (a) The importance of **natural selection** as a mechanism that accounts for evolution was reaffirmed. (b) The **chromosome/gene** theory was incorporated as a process or mechanism that causes evolution. (c) **Mutation variation** (changes caused when organisms are subjected to radiation), in addition to **sexual reproduction variation** are the two main causes that initiate evolution.

The Earth's Interior

The earth is a sphere with a diameter of 12,600 km and a circumference of 39,800 km. Actually, the earth is an oblate sphere because its polar diameter is slightly shorter than its equatorial diameter. Gravity (the earth's gravitational acceleration, g) is responsible for the earth's sphericity. The oblate shape is attributed to the earth's rotation on its axis and the resulting centrifugal force.

Direct evidence concerning the nature of the earth's interior is limited to boreholes (the deepest of which is about 12 km), erosion (exposing deeper rocks), and volcanic eruptions (bringing deeper material to the surface). Indirect evidence which gives us information about the structure of the earth's interior from the surface to the center, is provided by seismic or earthquake waves. These seismic waves (which may also be reproduced by any explosion near the surface) travel through the interior solid rocks with a velocity that varies depending on the rock density, and are reflected back to the surface where they are recorded by seismographs.

Based on seismic waves, the earth's interior is divided into 4 concentric shells. These concentric shells, their composition, and their depth boundaries are listed below. In addition, the density of rocks is calculated to be up to 13 g/cm^3 in the inner core in comparison to an average of about 3 g/cm^3 of surface rocks, the temperature is estimated to be around $5,000^0$ C in the earth's center, and the pressure increases from one bar (atmosphere) on the surface to about 4 million bars in the center. The 4 concentric shells from the surface to the center:

1. The **Crust**: depth – Up to 70 km over the continents and up to 10 km over the oceans.

composition – over the continents, mainly **granite**, an igneous rock (cooling of magma) which is light colored (feldspar and quartz). Over the oceans, mainly black volcanic **basalt** (dark Fe and Mg minerals).

Note: The crust also includes a relatively very thin layer of loose sediments or sedimentary rocks which are deposited nearest to the surface and contain the fossils or the evidence for past life.

2. The **Mantle**: depth – Up to 2,900 km.

composition – all dark Fe and Mg minerals (**mafic and ultramafic rocks**).

3. The **Outer Core**: depth – Up to 5,100 km.

composition – **liquid metals. Primarily Fe** and some Ni.

1.   The **Inner Core**: depth – Up to 6,370 km.

composition – solid metals. Primarily Fe and some Ni.

We know that the outer core is liquid because one type of seismic waves, the S waves or shear waves, do not travel through the outer core since these waves do not propagate through liquids. Why is the outer core liquid? As we noted above, heat increases from the surface of the earth to its center. Heat increases with depth at an average rate of about 50°C per kilometer and is known as the geothermal gradient. Radioactivity, especially in the mantle, adds additional heat to the interior. Also as noted above, pressure or, in this case, overburden pressure (the weight of the rocks above) also increases with depth. Therefore, in the earth's interior, there is an interplay between heat, which tries to melt the solid rocks, and overburden pressure, which tries to keep the rocks solid. At the depth of the outer core, where there is also a change in composition (see above), heat wins or prevails over the overburden pressure and, therefore, makes the outer core liquid.

Plate Tectonics

Geologists constructed a model of **plate tectonics**, based on evidence for continental drift and sea-floor spreading (some of this evidence will be presented below). This model depicts the earth's surface (both land exposed or covered with water or ice) as subdivided into a number of distinct **lithospheric plates** (spherical rigid rock plates) that move relative to each other in spherical geometric fashion. These plates are part of the rigid lithosphere, the outermost part of the earth which includes the crust and part of the upper mantle, down to 100 kilometers thickness. The concentric 100 km thick lithosphere sits on top of the concentric 150 km thick **asthenosphere** (the "sick sphere") which is not rigid but plastic or viscous. Why plastic instead of solid?

As noted in the section above (the Earth's Interior), there is an interplay between interior heat, which tries to melt the solid rocks, and overburden pressure, which tries to keep the rocks solid. At the depth of the asthenosphere, heat wins or prevails somewhat over the overburden pressure and, therefore, makes the asthenosphere rocks plastic. It is believed that most of the magma or lava that travels towards the surface (the direction of least pressure) originates in the asthenosphere through the partial melting of some rocks in this area where there is a local increase of heat (radioactive hot spots) or decrease of pressure (e.g. surface erosion).

The plastic rock material in the asthenosphere moves slowly by convection, forming convection cells very much like the convection cells of boiling water. In places where the convection cells in the asthenosphere move upward, the rigid lithosphere above it arches upward. In places where the convection cells move laterally, they pull or drag along portions of the rigid lithosphere; much like floating wood is pulled or dragged by water waves.

In places where the lithospheric plates move apart (i.e. form diverging boundaries), they create the earth's mid-ocean ridges. At other places, the plates form converging boundaries or collide head on and form mountain ranges such as the Alps or Himalayas. At other converging boundaries, instead of a head collision, one plate is subducted under the other, forming the earth's deep ocean trenches and adjacent linear volcanic mountain belts such as the Andes Mountains of South America or the volcanic island arcs of the Circum Pacific. At still other converging boundaries of a circular earth, the plates do not collide but slip past each other as in the case of the San Andreas Fault in California. Where plate boundaries interact as noted above, there is usually tremendous volcanic and earthquake activity.

Supercontinent **Pangea** is the name given to one continuous landmass, (surrounded by **panthalassa** or one continuous ocean). Geologic evidence shows that the last Pangea (it's estimated that there is a supercontinent cycle that lasts 500 million years) formed at the end of the Permian Period and its break-up began in late Triassic or 200 million years ago. The break-up formed the present lithospheric plates such as North America, South America, Africa, Eurasia and India-Australia, all of whose boundaries include both continental and oceanic crust. Plate Antarctica includes practically continental crust only, and the Pacific plate includes practically oceanic crust only. These plates move relative to each other with a velocity that ranges from about one inch to a few inches per year depending on their position on the spherical globe.

The supercontinent idea was introduced in 1915 by Alfred Wegener, a German explorer/geophysicist. Many geologists and other scientists around the world labeled him as "mad" since they believed in the permanence of continents and ocean basins. However, many of these scientists eventually outlined evidence that helped support Wegener's idea and, furthermore, they incorporated his idea in the unifying theory of **Plate Tectonics** which they formulated in the 1960's. Briefly, here is a very small fraction of the important evidence they presented. This evidence makes Plate Tectonics as one of the most generally accepted theories in science.

a. Global Geographic evidence: If pushed together, the continents fit like a jigsaw puzzle. Present computer models show an astonishing fit, especially the one of South America and Africa. Also, corresponding mountain ranges across many continental margins (e.g. North America and Europe) are of the same age, composition or type, and structure.

b. Fossil Evidence: Some fossils are uniquely found only in certain corresponding areas of continents, and nowhere else in the world, and separated by thousands of kilometers of ocean. The organisms represented by these fossils could not swim or fly across vast oceans. Their age is over 200 million years. The only way one can explain their existence in both places is to bring the continents together as they should have been part of **Pangea** at a time over 200 million years ago. Two of many examples: the Mesosaururs, a Triassic fresh water amphibian, found at the tip of South America and the tip of South Africa, and Lystrosaurus, a Triassic land reptile, found in northeastern Africa, western India, and northern Antarctica.

c. Oceanographic evidence: Before the 1960's, geologists believed in the permanence of continents and oceans. Oceanographers believed that if they could access the "smooth" ocean bottom, they would find sequences of sediments and fossils representing the entire history of the earth. In the late 1940's, oceanographers started using research vessels, dredging and coring, and generally mapping the ocean floor. What did they find? In the center of the oceans, mid-ocean ridges, the longest continuous mountain ranges on earth, above 3,000 meters high in some areas, with central grabens (rift valleys) and tensional fractures. At the margins of the oceans, they discovered deep trenches adjacent to volcanic island arcs or continental margins.

Also, the ocean crust was found to be made almost entirely of the volcanic rock basalt with relatively minor sediments. Fossils found in these sediments had a maximum age of 180 million years, hardly the entire history of life on earth. The ocean floor itself, made of basalt, was not older than about 180 million years.

All these findings are consistent with continental drift or sea-floor spreading and the plate tectonics model of a breaking Pangea separating and moving slowly from the divergent boundaries represented by the mid-ocean ridges and moving towards the convergent boundaries represented by deep trenches and volcanic island arcs. The maximum age of the fossils and that of the basalt (both about 180 my) is consistent with the break-up of Pangea at 200 my ago. At the rate of continental drift, it would take another 300 my before all continents meet on the other side of the globe to form a new Pangea. Also, it took a few million years (20 my) before the oceans opened wide enough to receive sediments and fossils. The oceanographers' big surprise? The present oceans represent some of the youngest major features on the surface of the earth. They expected the oceans to be close to the age of the earth (4,600 my).

d. Isotopic Dating of Ocean Floor Basalts: A very large number of ocean floor basalt samples (usually dredged from the ocean floor) were dated during the last fifty years. All these samples show that the youngest oceanic floor is near the mid-ocean ridges (close to the center of the oceans), and the oldest oceanic floor is furthest away from the ridges towards the trenches (close to the margin of the oceans). This age increase from the ridges to the trenches is symmetrical on both sides of the ridges. The oldest basalt, near the trenches is not older than about 180 million years, and the youngest basalt, near the ridges, is as young as the latest basalt eruption from the ridge center (graben or rift valley) which could be as old as last year's or yesterday's eruption.

This powerful evidence clearly shows that the ocean floor is spreading symmetrically away from the center of the ocean or the mid-ocean ridges and is moving towards the margin of the ocean or the trenches.

e. Paleoclimatic evidence: Late Permian to early Triassic glacial deposits (over 200 my old) are shown to be distributed in South Africa, in southern South America, western India, and southern Australia. The deduced direction of ice movement indicates that the glacial ice was radiating from South Africa towards South America, India, and Australia and was essentially moving over the oceans, an impossible situation since glaciers do not form or move over the oceans.

Here again we have significant evidence that supports continental drift because the direction of ice movement would make sense if we bring the above continents together and allow them to assume their positions in the supercontinent Pangea that was forming towards the end of the Permian. If we do this, then the ice would be moving over land instead of over the oceans.

f. Satellite (GPS) evidence: With this technology, we can measure the yearly continental drift by focusing on certain landmarks (e.g. the Empire State building) on the surface. The building is shown to drift westward (a few centimeters) only because of continental drift or, in other words, because the continent on which it was built is drifting westward.

GLOSSARY

(A page number after an entry refers to additional information or explanation in the Background Information section.)

A

Abiogenesis: Relating to the idea that life originates from nonliving matter.

Acanthodian: The first non-marine fish with jaws in the Silurian and Devonian periods.

Aerobic: Using or consuming free oxygen (O^2).

Alpine – Himalayan Orogenic Belt: Mountain building extending from the Alps in Western Europe to the Himalayas in Southeast Asia.

amino acids: A group of organic compounds, in the structure of proteins, that are essential to human metabolism.

amniote egg: The embryo in this egg develops in the fluid-filled amnion cavity.

Anaerobic: Organisms that do not need free oxygen (O^2) for respiration.

Anatosaurus: An Ornithischian herbivorous dinosaur genus with flattened bill-like mouth, also known as duck-billed dinosaurs.

Angiosperm: The flowering seed plants.

Archaeopteryx: The oldest fossil bird from Jurassic rocks in Germany. It has feathers and a   wishbone but it also retains some of its ancestral reptile (dinosaur) characteristics.

Archean Eon: The earliest period in the Earth's history, from 4,600 million years to 2,500 million years before the present. See the Geologic Time Scale on p. 71.

Australopithecus: This "human relatives" genus represents several extinct species that existed during the Quaternary period of the Cenozoic (4.2 to 1.2 million years ago).

Autotrophic: Describes organisms that synthesize their own food through the process of photosynthesis.

B

Basalt: A black, fine-grained igneous rock that forms through the cooling of lava on the surface of the earth. The lava reaches the surface through a volcano or a fissure. (p. 66)

Big Bang: A theory for the origin and evolution of the universe which began with the explosion of an infinitely small and inconceivably dense and hot point, resulting in an expanding, cooler and less dense universe.

Borates: The class of minerals that contains the $(BO_3)^{-3}$ anionic group. The most common of the minerals in this group is borax, a hydrous sodium (Na) borate.

Brachiopods: An invertebrate phylum that first appeared in the Cambrian as part of the skeletonized animals explosion. The brachiopods were benthonic, sessile suspension feeders.

Brachiosaurus: Saurischian, herbivorous, quadrupedal dinosaurs that attained giant dimensions, probably the largest land animals that ever lived.

C

Cambrian: The first (oldest) period in the Paleozoic Era. See Geologic Time Scale, p. 71.

cartilaginous fish: Fish with an internal skeleton of cartilage such as a shark.

Catalyst: Something acting as the stimulus in bringing about or hastening a result.

Cenozoic Era: The "new life" era that began 66 million years ago. See the Geologic Time Scale, p. 71.

Chordates: Animals of the phylum Chordata which possess a notochord, similar to the vertebral column of the vertebrates.

clay minerals: Usually the term "clay" defines fine-grained, earthy material that becomes plastic when mixed with a small amount of water. Clays usually consist mainly of clay minerals which are essentially hydrous aluminum silicates.

coelacanth: Primitive marine fishes considered possible ancestors to amphibians. They were known only in fossil form until a recent discovery (1938) of a "living fossil."

Columbia River Basalt: The river plateau built by overlapping fissure lava flows.

Comets: Orbiting members of the Solar System, numbering in the millions, with a luminous mass and a long luminous tale.

Constellation: A number of fixed stars arbitrarily considered as a group.

Corals: The stony skeletons secreted by certain marine polyps, sometimes forming massive coral reefs in tropical climates.

Core of the Earth: The innermost concentric shell of the Earth, the Inner Core, is made up of solid iron and nickel. The Outer Core above it is made up of solid iron and nickel. See the Earth's Interior, p. 76-78.

Craton: The stable nucleus of a continent consisting of Precambrian rocks.

Cretaceous: The last period of the Mesozoic Era. See Geologic Time Scale, p. 71.

Cro-Magnons: A member of our species (Homo Sapiens) that lived in Europe from 35,000 to 10,000 years ago.

Crossopterygians: Lobe-finned lung fish believed to be the ancestors of amphibians.

Crust: The Earth's outermost concentric layer. See the Earth's Interior, p. 76-78.

Cyanobacteria: Photosynthesizing prokaryotes known as blue-green algae.

Cynodonts: Advanced mammal-like reptiles believed to be the ancestors of mammals.

D

Devonian: The Paleozoic period known as the "age of fish." See Geologic Time Scale, p. 71.

Dinosaurs: Mesozoic reptiles that dominated the Earth for about 140 million years.

DNA molecule: Nucleic acids that serve as the blueprint for reproduction of all eukaryotes.

E

Ediacaran faunas: Late Proterozoic animal fossils found in Australia.

Endothermic: Vertebrate warm-blooded animals that maintain their body temperature within narrow limits by internal processes (opposite of ectothermic).

Enzymes: Protein-like substances that act as organic catalysts in initiating or speeding up certain chemical reactions.

Eukaryotes: Organisms whose cell(s) has a nucleus that contains the DNA blueprint for reproduction.

Exoskeleton: An exterior hard shell that provides protection for an organism.

F

Flowerless Seed Plants: Plants that produce seeds without flowering. These plants are officially classified as gymnosperms, from the Greek meaning "naked seed plants."

Flowering Plants: Plants that produce seeds through flowering. The plants are officially classified as angiosperms.

Forams: Short for foraminifera which represent tiny one-celled floating sea animals with calcareous shells.

Fossil: Remains or traces of once-living pre-historic organisms that are preserved in rocks.

G

galaxy: Any of innumerable large grouping of stars (often billions of stars).

gametophyte: Gamet-bearing plant generation that reproduces by eggs and sperms.

Geologic Time Scale: A chart listing the designation of the earliest geologic time at the bottom followed upward by progressively younger time designations, p. 71.

glaciation: The origin, expansion, and retreat of glaciers as well as their impact on the earth's surface.

glacier: A mass of ice that moves or flows over land under its own weight.

God particle: Informal name for Higgs boson.

gymnosperms: Flowerless seed plants or "naked seed plants."

H

Hadron collider: The world's largest and most powerful particle accelerator that started up in September of 2008.

heterotrophic: Organisms that do not produce their own food but feed on other organisms.

Higgs boson: A hypothetical, massive subatomic particle which is obtained after colliding two Hydrogen atoms travelling at the speed of light. Its existence (confirmed in 2013) would explain the masses of the elementary particles. (God's particle is its informal name)

Higgs field: Invisible energy field that exists everywhere in the universe and is accompanied by a fundamental particle called the Higgs boson.

Himalayan Orogenic Belt: See Alpine-Himalayan Orogenic Belt.

hydrothermal vent systems: Warm water systems near mid-ocean ridges providing the energy that is needed to form the life-forming amino acids.

Hominids family: The family of bipedal primates that includes our own genus Homo.

Hominoids superfamily: It includes the family of Apes and the family of Hominids, the human relatives and humans.

Homo: The genus of hominids which is made up of our own species, the Homo sapiens, and their ancestors or human relatives, the Homo erectus and the Homo habilis.

I

Iapetus Ocean: A Paleozoic ocean between North America and Europe (in the position of the North Atlantic ocean) that closed during the Late Paleozoic through continental collision.

Ice Age: The period of extensive continental glaciation between 2 million and 10,000 years ago.

ichthyosaur: The Mesozoic marine reptiles that resemble porpoises.

isotope: Two or more forms of an element having overall the same chemical properties and the same atomic number but different atomic weights (or mass numbers).

J

jovian planets: The four outer planets (Jupiter, Saturn, Uranus, and Neptune) that are relatively large, have low density, and are mostly gaseous with rocky cores.

Jurassic: The middle period of the Mesozoic Era. See Geologic Time Scale, p. 71.

L

labyrinthodont: Devonian to Triassic amphibians believed to be the ancestors of reptiles.

Laurentia: A Proterozoic continent composed primarily of North America, Greenland, and parts of northern Scotland and Scandinavia.

Laramide orogeny: Late Cretaceous to Early Cenozoic phase of the northern part of the Cordilleran orogeny that formed many structural features of present-day Rocky Mountains.

lung fish: fish with lungs as well as gills.

M

magnetic field: A region of space where magnetic substances are affected by an appreciable magnetic force emanating from the Earth.

mantle: The thick concentric layer of the Earth's interior, between the crust and the core. See the Earth's Interior, p. 76-78.

mammals: A class of warm-blooded, usually hairy vertebrates whose offspring are born live and fully developed and are taken care of by the parents.

marsupials: Pouched mammals, such as kangaroos, that give birth to their offspring in a very immature state that requires further development in the pouch.

mass extinction: Rates of extinctions greatly accelerated as in the case at the end of the Paleozoic or the case at the end of the Cenozoic.

meteorite: That part of a relatively large meteoroid or asteroid that survives passage through the atmosphere and falls on the Earth's surface as a mass of metal or rock.

Meteorite Impact hypothesis: The idea that a large meteorite, about 10 km in diameter, impacted the Earth and created a lot of dust and ashes which blocked the Sun's energy and caused mass extinctions of the dinosaurs and other species.

Mesozoic Era: The "middle life" era between the older Paleozoic Era and the younger Cenozoic Era. See the Geologic Time Scale, p. 71.

mid-ocean ridge: A major elevated linear feature (mountain range) of the sea-floor. It occurs at a divergent plate boundary. See Plate Tectonics p. 79-84.

Milankovitch theory: Explanation of the causes of ice ages as relating to irregularities in the Earth's rotation tilt and the geometry of its orbit.

mineral: A naturally occurring, inorganic, solid that has a definite chemical composition and a characteristic internal structure. (p. 65)

Mississippian: A Late Paleozoic period. See Geologic Time Scale, p. 71.

Mitochondrial DNA: Resilient genes inherited only from the mother and are, therefore, useful for tracing family trees.

mollusks: A large phylum of invertebrate animals including the oysters, clams, snails, squids, etc.

molybdates: A class of minerals of the anionic group $(MoO_4)^{-2}$ such as wulfenite $PbMoO_4$.

monomers: Simple organic compounds, such as amino acids.

monotremes: The egg-laying mammals such as the platypus of Australia.

mosaic evolution: Evolutionary concept relating to organisms with features retained from their ancestors as well as more recently evolved features.

N

Neanderthals: The oldest member of our species (Homo sapiens) that lived in Europe and the Middle East from 200,000 to 30,000 years ago. Recent evidence suggests that they may represent a parallel or separate species (Homo neanderthalensis).

O

Ordovician: The second oldest period of the Paleozoic Era. See Geologic Time Scale, p. 71.

Ornithischia: One of the two orders of dinosaurs (the other is Saurischia) characterized by a bird-like pelvis.

orogeny: An episode of mountain building usually involving several geologic processes.

ostracoderm: The "bony skinned" and jawless fish representing the oldest known vertebrates (Late Cambrian).

outgassing: The process that formed an early atmosphere through the release of gases from the Earth's interior by volcanism.

ozone layer: An appreciable concentration of ozone gas (O_3), high in the atmosphere (up to 50 km), which absorbs harmful UV radiation from the Sun.

P

paleontologist: The scientist who studies ancient life forms and their evolution through the use of fossils.

Paleozoic Era: The "ancient life" era that began around 570 million years before the present and ended around 245 million years ago. See the Geologic Time Scale, p. 71.

Pangea: The supercontinent (Greek for "one land mass") consisting of all of Earth's landmasses that formed at the end of the Paleozoic Era. The name was proposed by Alfred Wegener.

pelycosaurs: The fin-back or sail-back reptiles that thrived during the Permian period of the Paleozoic Era.

Pennsylvanian: The Late Paleozoic period. See Geologic Time Scale, p. 71.

Permian: The last period of the Paleozoic Era. See Geologic Time Scale, p. 71.

photosynthesis: The autotrophic process of some organisms (algae or plants) that use the Sun's radiant energy plus chlorophyll to combine CO_2 plus H_2O to produce food (hydrocarbons) plus oxygen (O_2).

placental mammal: A mammal with a placenta to nourish the developing embryo.

placoderm: A marine "plate-armored " fish with jaws.

Plate Tectonics theory: The theory that holds that segments of the Earth's lithosphere (upper 100 km thickness) move relative to each other.

Pleistocene glaciation: Ice ages of the last 2 million years of the Earth's history.

plesiosaurs: A Mesozoic marine reptile.

polymers: When simple organic molecules or monomers link together, they form polymers or relatively complex organic molecules.

Precambrian: The oldest era of the Earth's history ranging from the beginning of the Paleozoic Cambrian period or 570 million years before the present to the origin of the Earth or 4,600 million years before the present. See the Geologic Time Scale, p. 71.

primates: Mammals that belong to the order Primates which includes apes and humans.

prokaryotic cell: The cell that lacks a nucleus and therefore does not contain RNA or DNA which contain the blueprint for reproduction. It is the cell of bacteria or cyanobacteria.

Prosimians suborder: A suborder of the Primates order; it includes lemurs, tersiers and tree shrews.

Proterozoic: An eon of the Precambrian ranging between 2,500 and 570 million years BP. See the Geologic Time Scale, p. 71.

pterosaurs: The Mesozoic flying reptiles.

Q

Quaternary: The last period of the Cenozoic Era, ranging from the present to about 1.8 million years ago. See the Geologic Time Scale, p. 71.

R

radioactive decay: The spontaneous decay of an atom by emitting particles from its nucleus to eventually form a stable atom of a different element.

red dwarf: A star that is cooler, smaller, and of less luminosity than our Sun.

RNA: Ribonucleic acid, the blueprint for reproduction in bacteria.

S

Saurischia: The dinosaur order which is characterized by a lizard-like pelvis.

seedless vascular plants: Plants that reproduce in water by spores rather than seeds.

Silurian: The third oldest period of the Paleozoic Era. See Geologic Time Scale, p. 71.

solar nebula theory: A theory for the evolution of the solar system by the rotation and contracting of a cloud of gas.

Stromatolites: A biogenic structure produced by colonies of photosynthesizing bacteria or cyanobacteria (blue-green algae).

Solar System: Our Solar System includes: 8 major planets and one dwarf planet that orbit around the Sun; plus over 100 moons or satellites that surround the planets, plus a large number of asteroids, comets, meteorites, dust, and gases.

spore: Any small organism or cell that can develop into a new individual.

sporophyte: The spore-bearing plant that reproduces by spores.

stem reptiles: The Late Paleozoic reptiles from which originated all other land animals.

supercontinent: See Pangea.

supersymmetry: The origin of the cosmos theory that proposes that each of the 17 particles already identified has a nearly identical cousin in the shadows of the universe.

T

Taconic orogeny: Paleozoic Ordovician mountain building. See the Geologic Time Scale, p. 71.

terrestrial planets: The four innermost planets in our Solar System (Mercury, Venus, Earth, and Mars), composed of rock and metallic elements (Fe and Ni).

Tertiary: The first period of the Cenozoic Era. See the Geologic Time Scale, p. 71.

thecodonts: The dinosaurs, the pterosaurs, the marine reptiles and other reptiles, all evolved from the thecodonts (teeth in individual sockets) during the Triassic.

therapsids: Permian to Triassic mammal-like reptiles believed to be the ancestors of mammals especially the group known as the cynodonts.

Triassic period: The first period of the Mesozoic Era. See the Geologic Time Scale, p. 71.

Tyrannosaurus Rex: Belongs to the Saurischian order of dinosaurs, representing a genus that is believed to have been one of the largest carnivorous dinosaurs.

U

ultraviolet radiation: Light rays of about 4,000 angstroms wavelength, lying just beyond the violet end of the visible spectrum.

V

vascular plants: See "seedless vascular plants."

vertebrates: Animals that have a segmented vertebral column such as fish, reptiles, birds, and mammals.